Berdiyar Jolybekov

Impact of biointensive technologies on fertility of saline soils

Berdiyar Jolybekov

Impact of biointensive technologies on fertility of saline soils

ScienciaScripts

Imprint

Cover image: www.ingimage.com

This book is a translation from the original published under ISBN 978-3-659-82360-2.

Publisher:
Sciencia Scripts
is a trademark of
Dodo Books Indian Ocean Ltd. and OmniScriptum S.R.L publishing group

120 High Road, East Finchley, London, N2 9ED, United Kingdom
Str. Armeneasca 28/1, office 1, Chisinau MD-2012, Republic of Moldova, Europe
Managing Directors: Ieva Konstantinova, Victoria Ursu
info@omniscriptum.com

Printed at: see last page
ISBN: 978-620-8-51317-7

Table of contents

Introduction

Relevance of the topic. Preservation and increase of soil fertility, improvement of ameliorative condition of saline irrigated soils, as well as economical use of irrigation water, especially under conditions of their deficit, are the main problems of modern agricultural science.

Beginning from 2000, due to the low water level, agriculture started to decline. Only wheat-58.3; vegetables-0.5; melons-0.3 thousand hectares were sown. There was a huge damage to the environment, social and economic life of the region. In addition, there are negative consequences that strongly affect the quality of agricultural products and people's health.

As a result of irrigation with mineralized water, irrigated soils on the territory of the Republic of Karakalpakstan are subjected to secondary salinization and salinization, as a result of which a huge area of cultivated land is out of agricultural rotation, which in turn requires the development of non-standard technologies to combat such negative phenomena.

One such method is biointensive technology, a special form of farming, agroculture that does not require the use of chemical fertilizers or pesticides to grow crops. It focuses on methods of providing the soil with all the necessary components. It allows you to create healthy soil solely from the resources of your own homestead. This technology provides more than 2 to 6 times more yield as compared to conventional farming. Bio-intensive method will reduce water rates by 70-80%, fertilizers from 50% to 100%, and energy by 99% (Jevons,1993).

In connection with the above-mentioned, the issue of application of various water-saving irrigation methods and biointensive technologies in the Amudarya lower reaches is of paramount importance. This is conditioned by acute shortage of irrigation water and high cost of mineral fertilizers, comprehensive and negative impact of irrigation water quality on ameliorative state of irrigated soils and their fertility.

The annual volume of manure and other forms of organic fertilizers in the Republic of Uzbekistan is approximately no more than 8 tons for each hectare of irrigated land. This is insufficient for obtaining high yields of vegetable and other crops.

Search and creation of new water-saving and bio-intensive technologies, application of which will contribute to economical use of irrigation water, increase of fertility by improving soil ameliorative state and increase of yield of cultivated vegetable crops, will be a significant step in solving the above-mentioned problem.

A number of studies abroad on the application of biointensive methods have been conducted. The results of field trials have shown that vegetable crops at application of biointensive technologies grow twice as fast as the control, the yield increases due to the improvement of agrochemical and agrophysical properties of soil.

In Uzbekistan, the production of compost from household waste using the D. Jevons method was started by some researchers after 1995 in some homestead and dacha plots.

At present, there is some positive experience in the use of water-saving and biointensive technology together and separately. However, many issues, such as irrigation methods, soil treatment, obtaining compost from household waste, use of plastic bottle and soaking hose as an irrigator are poorly developed.

Therefore, the study on revealing the influence of different irrigation methods and biointensive techniques on fertility of irrigated meadow alluvial saline soils is relevant and is of scientific and practical interest. In connection with the above, the issue of application of various water-saving irrigation methods and biointensive technologies in the Amudarya lower reaches under conditions of acute shortage of irrigation water and its poor quality, as well as the high cost of mineral fertilizers, comprehensive and negative impact of irrigation water quality on the ameliorative state of irrigated soils and their fertility is of paramount practical and scientific importance, which determines the relevance of this work.

Purpose and objectives of the study. The main objective of this work is to study the impact of different irrigation methods and biointensive technologies on fertility of irrigated saline soils in the lower reaches of the Amu Darya, as well as the development of a water-saving irrigation system from domestic wastes, allowing to increase the productivity of saline irrigated soils and yields of vegetable crops.

In order to achieve the objective, the following tasks had to be accomplished:

- study the current chemical status of irrigation and drinking water in the Amudarya downstream;
- reveal the influence of different irrigation methods and biointensive technology on the content of water-soluble salts, organic matter and fertility of meadow alluvial irrigated soils subject to salinization;
- study the effect of different irrigation methods and biointensive technology on tomato and eggplant yields;
- Develop bio-intensive and water-saving technologies to increase productivity irrigated saline soils and vegetable crops.

Scientific novelty and practical value of the work - is that for the first time on irrigated meadow alluvial saline soils revealed the effectiveness of water-saving and biointensive technology developed by the author (the use of household waste on the basis of plastic bottles in the irrigation system) having no analogues, both in our country and abroad.

The effect of different irrigation methods on ameliorative state, soil fertility and yield of vegetable crops has been determined. On the basis of the obtained experimental data, scientifically substantiated conclusions and proposals for production were made.

Chapter 1. Current state of study of fertility of irrigated meadow alluvial soils of the Amudarya lower reaches (literature review)

The study of soil fertility, its reproduction and estimation of humus content were at different times studied by N.I. Zimina (1955,1961); K.A.Kachinsky (1956,1963); N.K. Balyabo (1961); J.Y. Batkaev (1962), J. Cook (1970); V.G. Loshakov, S.F. Ivanova (1975); L.L. Shishov, D.N. Durmanov (1985); M.V. Mukhamedjanov (1986).

Plant residues are a complete source of organic matter accumulation. It is fresh organic residues that greatly influence effective fertility (Kononova, 1963).

The data obtained by scientists prove that green fertilizer does not accumulate, but maintains the humus content in the soil at the initial level.

Data from experiments conducted in conditions of Belarus (Dovan, 1990) shows that leguminous siderates significantly increase the nitrogen content in the soil. Sideration cannot be considered unilaterally, i.e. only as a source of nitrogen fertilizer or organic matter, it has a complex effect.

According to E.K. Alekseev (1959), D.N. Prokopov (1976), and V.F. Trushina (1969), a highly effective method of increasing soil fertility is also stocking of green fertilizer (siderate), the use of which is possible both in main and intermediate crops.

In the experiments of Kusselevski Z. (1973) on long-term manure-fertilized plots the carbon content of organic compounds in soil increased.

According to T.V. Tyulin (1960), with increasing humus content in the soil, there is a tendency to increase the number of water-stable aggregates in the soil.

Under the influence of fertilizer application, proper agrotechnics, crop rotations, humus content in soils increases (Ryzhov, Ziyamukhamedov, 1971; Fursov, 1966).

M.M. Kononova (1960,1963) attaches exceptional importance in soil fertility depending on soil structure and nature of humus substances. Soil organic matter plays a complex and diverse role in the formation and development of soil fertility. Organic matter determines to a great extent the direction of soil formation processes, biochemical, chemical and physical properties of soils.

Annual cyclic processes of synthesis and slow mineralization of fresh organic matter and humus lead to systematic supply to soil solutions in the form of available mineral compounds of the most important elements of plant nutrition: nitrogen, phosphorus, sulfur, potassium, calcium, trace elements.

Organic substances influence the acidity and redox potential of soils. They are regulators of mineral nutrition of plants. Soil-inhabiting macro- and microorganisms draw energy and mineral compounds necessary for their existence from soil organic matter.

Soil organic matter is a universal accumulator and economical distributor of the most valuable part of energy necessary for normal exchange and turnover of substances in nature.

Therefore, the preservation and accumulation of effective forms of organic matter in the arable and subsoil horizons of soils through the application of organic fertilizers and grass sowing is an urgent issue.

According to studies by Minashina N.G. (1974), the greatest increase in humus occurred in arable and subsoil layer up to 40 cm thick. Humus content in 0-20 cm layer increased 3 times, on average for all soils, including light irrigated soils on rashes.

Humus content in soils depends on many factors: on the age of development, on the mechanical composition of soils, on hydrogeological conditions, on the intensity of irrigated farming, on the level of culture, on the type of agricultural plants, on the type of fertilizers used, etc.

The process of humus accumulation has a certain stability for each of the conditions of moistening and use. Within the arable layer the process of organic matter transformation proceeds with different speed and has time to manifest itself during one season (Minashina, 1974).

In the works of L.S. Baimatova and L.N. Slesareva (1990) it was established that against the background of plowing alfalfa to a depth of 40 cm the reduction of clumpy fraction and increase of agronomic valuable aggregates amounted to 7.9 and 3.8 %, respectively.

Anthropogenic pollution of the environment is currently on a large scale (Zhollybekov 1995, 1997, 2000, Jevons 1993, Zhollybekov 1996, Richardson 1988, Jollibekov 2005).

Irrigated agriculture in the Republic of Uzbekistan has been developing recently due to introduction of new lands into agricultural turnover. In this regard, in Karakalpakstan the area of irrigated lands was brought to 501.9 thousand hectares. As a result, production of raw cotton, rice and other agricultural products has increased dramatically (Zhollybekov, 1996).

Since 2000, due to the collapsing low water, the decline in agriculture began. There has been a huge damage to the environment, social and economic life of the region and moral trauma to the population. In addition, negative consequences appeared, which strongly affect the quality of agricultural products and the health of people (Zholybekov 2000).

As a result of irrigation with mineralized water, irrigated soils on the territory of the Republic of Karakalpakstan are subjected to secondary salinization. As a result, a huge area of cultivated land becomes unfertile and goes out of agricultural rotation (Zhollibekov 1997).

Over the last 50 years, the territory of the Southern Priaralie due to the increase of hardness, mineralization and amount of water-soluble salts in the waters of the Amu Darya River, irrigated soils everywhere became saline (Jolybekov 1995). As a result, only 111,600 tons of raw cotton and 1,058 tons of raw rice were produced in 2001, which in turn led to mass unemployment of the population.

Few plant residues remain in anthropogenic soils every year, and this leads to a gradual reduction of humus and nutrient elements in them. Taking this into account, it is necessary to develop a system of organic fertilizer application (Sattarov et al. 2001).

In Surkhandarya, Kashkadarya, Bukhara and Navoi provinces, as well as in the Republic of Karakalpakstan, since 1987-1988, there has been a decline in the productivity of irrigated lands and irrigation water. This is evidenced by the gradual decline in yields of all crops (Yakubov and Kudratov, 2001).

In conditions of Karakalpakstan with application of cotton-alfalfa crop rotation, in arable layer of soil, due to accumulation of large mass of small roots, humus content increased by 2.8-9.3 %. With additional application of manure at the rate of 40 t/ha, under the 4th year of cotton after plowing alfalfa, the humus content in the soil increased by 15.8-18.2 % (Ismailov, Zhollibekov, 2003).

The issue of preservation and increase of soil humus is acute in conditions of Karakalpakstan, where dryness and high temperatures, scarcity of vegetation limits humus formation.

Biointensive technology is a special form of organic agroculture that does not require the use of chemical fertilizers or pesticides to grow crops. It focuses on methods of providing the soil with everything it needs for its health that can be fully sustainable. Biointensive technology creates healthy soil solely from the resources of your own homestead. This technology provides 2 to 6 times more yield than mechanized chemical farming. The biointensive method will reduce water rates by 70-80%, fertilizer rates by 50% and up to 100%, and energy rates by 99% (Jevons 1993).

In China, agriculture was "miniaturized" in a similar way over 4,000 years ago. Crops were grown there for several millennia, densely spacing plants, but maintaining the fertility of the soil and not destroying its resources. In 1890, this allowed Chinese peasants to grow a complete diet for one person on an area of 540 square meters to 670 square meters, including the products of the then current animal husbandry.

Miniature forms of agriculture were developed, independently, 2000 years ago in various countries - Greece, Bolivia, Peru, Nepal, Guatemala, Mexico and Japan.

In the first, the French began to apply the intensive method, using only organic fertilizers. Biodynamic and French intensive farming methods emerged in Europe in the late twentieth and early twentieth centuries. French, intensive methods of cultivation were tested in the 1890s on a plot of about one hectare of land near Paris. Crops were grown on horse manure about half a meter thick. The plants were grown very close to each other, so that when they grew, their leaves touched each other. This created a microclimate and a layer of living plant mulch, which inhibited the growth of weeds and stored moisture in the soil.

In the early 1920s, biodynamic methods were developed by Rudolf Steiner, a brilliant Austrian thinker, philosopher and educator. He noticed a decrease in the nutritional value of crops and some reduction in yields and concluded that this was due to the use of chemical fertilizers and pesticides. He found that at the same time the number of crops affected by disease and insect pests increased. Even the ancient Greeks, two thousand years ago, noticed that plants grow particularly well on landslides. Air, moisture, heat, nutrients and roots easily penetrate the loose soil. Farmers using the biointensive methodology make raised beds with a shape similar to that of landslides. At this time farmers and gardeners make rows 15 cm wide with a large inter-row space. It is more difficult for plants to develop in such rows because of excessive air penetration into the soil and because of large fluctuations in its temperature and humidity. Thus, when watering, water floods the rows, roots are in the water. And there is a washout of soil from the rows and from the upper roots. As a result, the beneficial microbiological environment around the roots, necessary to prevent disease and to prevent nutrients in forms digestible by plants, is disturbed, and pests can even multiply (About 75% of the beneficial microbiological environment is concentrated in the top 15 cm of the soil layer). When water goes deep into the soil, the upper layers dry out and microbial activity is greatly reduced. At the same time, the plant rows are subjected to large temperature fluctuations.

Between the 1920s and 1960s, Alan Chadwick combined the biodynamic method with the French method of intensive gardening into a single biodynamic/French, intensive method. In the United States, this method became known in the sixties when Chadwick applied it to a student garden on the campus of the University of California, Santa Cruz, on a four-acre plot.

Based on Alan Chadwick's Biointensive Gardening Method, the book "How to Grow More Vegetables..." (D. Jevons, 1993) details how to grow vegetables for a family of four on 75 square meters or less, using only organic fertilizer, while maintaining a nutritious diet. (D. Jevons,1993) details how to grow vegetables for a family of four on 75 square meters or less, using only organic fertilizers while preserving the soil nutrients of future crops, and with minimal water and labor. Over the last few years, Ecology Action's staff has been developing and simplifying the Biodynamic and French intensive methodology, and its name has changed. It is referred to simply as "Biointensive Methodology" of vegetable gardening or "Biointensive Mini Farming". In 1972, the Ecology Action Group set up a biointensive vegetable garden in the city of Palo Alto to conduct research, show what had been accomplished, and teach the methodology used.

Healthy soil produces healthy plants that make healthy people who can build a healthy ecosystem and planet.

Natural natural systems are sustainable and do not require additional resources from the outside, whereas human-made agro-systems cannot function without bringing in additional resources, huge inputs of manual and mechanized labor. No culture or nation can develop sustainably without the judicious use of resources that provide for daily human needs and production. Similarly, it is impossible to survive without sustainable agriculture that provides food.

After the 1960s, the southern part of Priaralie turned into a region of accumulation of all negative consequences of inept human economic activity. Intensive development of reclamation-unfavorable lands, discharge of polluted collector-drainage water into the river, long-term monoculture of cotton and rice, lack of profitable economic mechanisms for saving irrigation water and improving soil fertility have led to large-scale deterioration of reclamation state of irrigated soils (Zholybekova, 1994, 1995, 1999, 2000).

Biointensive Mini Farming (BIMF) is an organic system. Synthetic granular and liquid pesticides and other commercial fertilizers are not used. Its organic nature means less environmental pollution and greater stability of the whole agroecosystem in which natural biological control elements proliferate.

The beds and the paths between them are made permanent. The farmer does not have to walk on the finished beds, which reduces soil compaction. Similarly, fertilizer and water are not wasted on the paths and they do not have to be dug up unnecessarily.

The beds are fertilized with plenty of compost and other natural fertilizers as needed. Appropriate production and use of compost is most important in maintaining soil fertility. BIMZ allows the use of manure, but relies primarily on plant-based compost. Proponents of the system believe that over-reliance on manure application creates nutrient imbalances (Jevons 1997).

On the other hand, a variety of vegetation can be grown right on site. Even in urban areas, plant materials such as leaves, grass clippings and kitchen scraps are widely available.

Double digging and concentration of fertilizer in the beds ensures high reproduction of crops. Additional crops are used to expand space and improve pest and fertility control.

The use of BIMZ results in low losses. Nutrients are recycled through composting of all crop residues. Cover crops, crop diversity and no plowing of the land ensures low soil and nutrient losses to erosion and depletion.

High financial investments are the most significant barriers in modern agriculture. BIMZ uses small areas, does not require expensive irrigation equipment, purchase of fuel and lubricants and agricultural machinery, and reduces annual costs. As a result, it is accessible to people with limited finances and land resources.

However, the use of a closed-loop system is not necessary if other sources of organic materials are available: manure, sawdust, food waste, etc. The above also applies to forage production in mini-pastoralism.

In the process, the study developed bio-intensive technologies to improve soil fertility and safe methods of utilization of domestic waste, tree and plant residues.

BIMZ is characterized by two advantages that no other farming system has.

First, less stress on the soil compared to mechanized methods.

Secondly, the lowest cost in terms of capital investment. For very small farms (mini-farms), this method is not only economically viable but also superior to any other method.

Therefore, we decided to use bio-intensive technologies to increase soil fertility in the epicenter of the Aral ecological crisis.

N.I. Toshmatov (1996) suggested that for the preparation of compost from industrial waste, the chemical composition of the waste should first be determined and the amount of heavy metals in it should not exceed the MPC.

T.S. Zakirov, M.M. Tashkuziev et al. (2000) revealed that humification processes under alfalfa are more intensive than under cotton and humus accumulation increases from the application of organic and mineral fertilizers.

M.M. Mirzaev, M. Mirzaev (2000) proved that the rate of nitrogen fertilizer application can be halved when using siderates, especially legumes.

L.N. Shchapova (2000) on the basis of micro plot experience found that the annual application of peat, manure and straw contributed to the accumulation of humus in all periods of observation and proved that the annual application of organic fertilizers at a dose of 6-16 t/ha is more effective than high doses for a long period.

TS.D-TS. Korsunova (2000) made the following conclusions on the basis of vegetation field experiments: manure is the best source of humus substances formation; alfalfa residues lead to increase of total organic matter.

A.Vakhobov, O.Mirzaev, Z.Zhuraev (2001) studied the effect of biohumus on growth and development of cotton in field experiments in Fergana valley. With the use of biohumus cotton yield increased by 19.7 c/ha and 30 days earlier ripening in comparison with the control variant.

O. T. Karimov (2001, 2002) found that the content of mineral forms of nitrogen, mobile phosphorus and exchangeable potassium in soils increases when organic fertilizers are applied separately and together with mineral fertilizers. Application of bio-humus separately or together with mineral fertilizers has a favorable effect on cotton growth and development.

A.I. Razhabov, N. Gafurova (2001) found that in irrigated typical sierozem low-mobile trace elements accumulate in the upper horizons of soil profile as a result of biogenic accumulation and are equalized by overgrazing.

K.D. Davranov, D.H. Karshieva (2002) under the term biological farming implies a system of agrotechnical measures that ensure the orientation of microbiological processes along the path of expanded reproduction of soil fertility, increasing the sustainability of yields and quality of products, reducing environmental pollution.

H.H. Imamova, H.T. Sharipova et al. (2002) noted that in compost from manure, the number of ammonificators, bacteria using mineral nitrogen decreases, and in compost with poultry manure, the number of bacteria growing on mineral nitrogen and oligotrophs increases, indicating that the compost is enriched with protein nitrogen.

S.Ya. Bezdnina (2000) analyzing the results of numerical experiment on prediction of water-salt regime of ordinary chernozems of High Syrtes came to the conclusion that with deterioration of irrigation water quality even in the presence of leaching water regime, the degree of danger of development of general salinization processes, sodium and, especially, magnesium salinization increases at low degree of initial salinization.

V.E. Prikhodko (2000) revealed that reversible, controllable changes under irrigation are: content of humus, water resistant aggregates, gypsum, easily soluble salts, nutrients, groundwater level. Irreversible degradation processes are found to be: destruction of microstructure, unfavorable and transformational changes of organic matter (decrease in the number of its reactive microforms, increase in dispersion, separation of coagulated and dispersed microforms) and mineral plasma, leading to weakening of connection between them.

D.S. Sattarov, O.R. Kazak (2000) revealed that the application of mineral fertilizers alone promotes the active use of soil microelements, as a consequence, there is a decrease in their content. When soils are used in agricultural production without organic fertilizers there is a decrease in humus content.

R.K.Kuziev (2000) studying the current ecological and reclamation state of irrigated soils proved that there is a tendency to decrease soil fertility in the Republic of Uzbekistan.

Turapov I. (1994), Khalikulov Sh.T. (1996), Kurvantaev R. (2000) notes that polyethylene film as a mulching material during the vegetation period differently affects the change of soil water regime. At the beginning of vegetation, soil moisture increases under the film, in the middle the positive effect of the film on moisture conservation decreases, and at the end the amount of water in the soil is equalized with the control.

Kurvantayev R., Khalberdiyeva M.R. (2001) revealed that at mulching of soil at ridge method of sowing by polyethylene film and organic mulches due to optimization of water properties favorable conditions for plant development are created.

Many scientists have studied the problem of irrigation in the lower reaches of the Amu Darya River. In 1924, V.V. Zinserling, having generalized the materials of researches of previous years, published information about lands suitable for irrigation and irrigation capacity of the Amudarya River with unregulated flow. In 1926, Engineer G.N. Vinogradov drew up the first scheme for utilization of water and land resources in the Kashkadarya river basin, which outlined the construction of reservoirs on the river and its tributaries, and also considered the possibility of a gravity canal for water supply from the Amudarya river (from Kelif) to the territory of Bukhara province. In 1924, under the leadership of F.P. Morgunenkov, research was carried out and "Scheme of development of irrigation systems in Southern Khorezm" was drawn up, in 1926 - "Scheme of Amu Darya river water transfer along the Kelif Uzboy for irrigation of lands in the deltas of Murghab and Tejen". In 1928-1930, together with V.V. In 1928-1930, together with V.V. Poslavskiy, E.M. Benyaminovich and others, he developed a project for reorganization of irrigation systems and collectors in the lower reaches of the Amudarya River. In 1930, under the leadership of F.P. Morgunenkov, the Central Asian Water Management Department drew up the "Scheme of land and water resources utilization in the Amudarya River". It shows for the first time possible sizes of irrigation (up to 4 mln ha) under partial regulation of Amudarya river flow. Amu Darya. In 1954, the "Sredazgiprovodkhlopok" Design Institute developed the "General Scheme of Water and Land Resources Use in the Amudarya River Basin" (compiled by S.A. Bayukov). In it the scales of irrigation development on domestic and regulated flow were defined step by step.

The middle and lower reaches of the Amu Darya River are characterized by irrigation of huge areas, often significantly distant from water sources. This is due to the construction of large river hydrosystems and canals of long length, often with cascades of pumping stations.

There are 4207.9 thousand ha under agricultural land in Karakalpakstan. Irrigated lands are partially used under perennial plantations, under sowing of technical and grain crops, homestead plots, etc. In 1975. 124,000 ha of irrigated land was occupied by cotton, 52,800 ha by cereal crops (of which 42,200 ha by rice and 8,800 ha by corn for grain), 8,500 ha by potatoes and vegetable and melon crops, and 63,100 ha by fodder crops. Uzgiprozem recommended average annual irrigation rates of 25000-27000 m^3/ha for rice crops on channel sediments and 22000-25000 for inter-channel depressions in the northern regions of Karakalpakstan. For cotton crops he recommended average annual irrigation norms - 12000 m^3/ha of which 6000m^3/ha leaching norm. In the Republic of Karakalpakstan leaching period is from September 21 to April 20. For slightly saline soils 1-2 leaching irrigations are carried out at the rate of 2000-4000 m^3/ha, for moderately saline soils-2-3 at the rate of 4000-6000m^3/ha, and for highly saline soils-3-4 at the rate of 6000-8000m^3/ha (Irrigation of Uzbekistan, 1979).

U.E. Ismailov (1983) found that the higher the salinity of irrigation water, the greater the amount of salts accumulated in the root zone.

In conditions of the Republic of Karakalpakstan, which is in the epicenter of ecological disaster on medium loamy, saline meadow soils with groundwater table of 1.5-2.2 m requires annual operational leaching at the rate of 3-4 thousand m^3/ha.

K.R. Nurzhanov (1983,1984) revealed that as irrigation rates increase, compaction increases and soil volume mass increases, especially in arable layer. Seasonal salt accumulation during vegetation period depends on irrigation and irrigation rates. He suggested (1984) to wash out salts from the root-inhabited layer of soil accumulated during the growing season, it is necessary to carry out annual operational leaching at the rate of 4.0-5.0 thousand m^3/ha. Kh.T.Riskieva, O.Toshkenboev (2002) on the basis of studying the change of humus state of meadow soils of desert zone under the influence of irrigation came to the conclusion that irrigation and agricultural use of hydromorphic soils of desert zone is accompanied by improvement of their chemical properties, increase of organic matter reserves and improvement of humus qualitative composition.

S.A.Abdullaev, A.J.Bairov (2002) suggest that in order to regulate balanced mineral nutrition of plants it is necessary to revise the system of fertilizer application for different crops taking into account the use of new forms of organic fertilizers and local mineral raw material resources. In the Republic of Uzbekistan, continuous and furrow irrigation with large irrigation rates were widely used.

Under Central Asian conditions, the following different types of irrigation are used in vegetable crops cultivation: Reserve (water-loading) irrigation is used in areas with low winter precipitation to increase the moisture content in the soil by the time of sowing or planting vegetable crops and reduce the number of vegetation irrigations.

The norms of reserve irrigation depending on the mechanical composition of soil and groundwater table vary from 1000 to 2500 m^3.

Pre-plowing irrigation is used to moisten the soil prior to winter or pre-sowing (summer) plowing at the rate of 1000-1200m^3/ha.

Provocative watering in the spring and summer period is carried out in order to induce the sprouting of weeds, which are then destroyed by cultivation or plowing the soil to a shallow depth.

Flush irrigation in the Republic of Uzbekistan is widely used on saline soils. Their rate varies from 1500-2000 to 6000-8000 m^3/ha depending on the degree of salinity and physical properties of soil. Soils are washed in late fall-winter in frost-free days or in early spring.

Pre-sowing or preplanting irrigation is carried out to moisten the soil before sowing or before planting vegetable seedlings. Watering by furrows with the rate of 400-600 m^3/ha.

Feeding irrigation is carried out after sowing or planting vegetables to ensure more friendly sprouting of sown seeds or better rooting of planted seedlings. Irrigation rate is 400-500m^3/ha.

Vegetation irrigations provide plants with moisture during growth and development. The number of vegetation irrigations and irrigation rates depend on soil and climatic conditions, biological features of the growing plant, the timing of its sowing (planting) and other factors.

Feeding irrigation of vegetable crops through inter-row irrigation is practiced almost in all vegetable farms of the Republic of Uzbekistan.

Sprinkling irrigation has significant advantages over furrow irrigation - there is no need for a shallow irrigation network, water consumption is significantly reduced. In Uzbekistan, sprinkling is recommended in hot summer months in addition to surface irrigation. Irrigation rates are 250-500 m^3/ha (Balashev and Zeman, 1981).

K. Ovezova (1991) under conditions of the Republic of Turkmenistan developed a highly effective water-saving technology of melon crops cultivation by centering-irrigation furrows with deep loosening of seed bed. In developed technology rational irrigation scheme 1-2-2 with irrigation rates 850-1490-1550 m^3/ha with leaching irrigation rates 2000 m^3/ha and total annual irrigation water consumption 5890 m^3/ha.

A. Shokirov (2001) in the conditions of Tashkent region put field experiments at different densities of eggplant. The most effective density of eggplant was 70 x 40 and yielded 83.0 c/ha.

New tomato varieties, along with such features as high yield, resistance to diseases, good taste and technological qualities, should be ecologically clean, transportable and have an attractive marketable appearance (Ermoleva, 2003). New tomato varieties developed in the conditions of the Republic of Uzbekistan yields on average from 35 (Sevara) to 65 c/ha (Surkhon-142).

The above literatures indicate the greater importance of humus in creating a water tolerant structure.

Thus, analyzing numerous literary data we can conclude that in the territory of the Republic of Karakalpakstan no scientific research on soil fertility improvement using biointensive and water-saving technology in vegetable crops cultivation has been conducted. This is the reason for the necessity of the present research.

Chapter 2. Objects and methods of research

The object of the study was meadow-alluvial irrigated saline soils. Irrigation water of Kyzketken canal and tap water.

2.1 Soils

In 2000, the northwestern part of Uzbekistan, especially the Karakalpak Republic, was subjected to severe drought and experienced water shortages. A significant part of rice, cotton, vegetables, fruits and other crops was lost. Improper use and operation of irrigation systems and unsustainable agricultural production have exacerbated the effects of the drought. Land degradation has already affected vegetable and fruit production.

The area of Nukus city is 120000 km^2, population 245000 people. On this area several hundreds of hectares of land suitable for agriculture are vacant, where it is possible to grow and obtain high ecologically clean yields of vegetable and melon crops and other crops using bio-intensive and water-saving technology - a new form of farming of urban population.

In order to characterize soils and to reveal the influence of biointensive technology on soil formation process, one transect-1 (September 16, 1999) before the experiment and transect-2 (September 15, 2004) after three years of research were laid down during the years of research.

Section No.1. RK, Nukus, 20 microdistrict, September 16, 1999. Nukus, 20 microdistrict, September 16, 1999. Initial state of soils of the experimental plot, fallow.

Soils of these sections are homogeneous by character of genetic horizons. By mechanical composition they are light loamy (Table 2.1.1). The higher content of true aggregates in the profile of section No.1 (fallow land) is apparently due to high hypsicization of soils of this section throughout its depth (see Table 2.1.1).

By the sum of salts and character of salinization the soil of experimental plots (section No.2) sharply differs from the initial state of these soils (section No.1)

By type of salinization soils of section No.1 (fallow land) belong to very strongly saline soils (solonchak). In the upper horizon dense residue is 6.505%, and at the depth of 80-120 cm. decreases to 3.315% by weight. In terms of salinity type, the upper (0-20 cm.) horizons are classified as sulfate-chloride and below to a depth of 80-120 cm. chloride-sulfate. (Table 2.1.2)

As a result of development, irrigation and application of biointensive farming methods these highly saline soils have almost completely desalinized. By degree of salinization the upper horizons (0-10 and 10-45 cm.) are characterized as slightly saline, horizons 45-60 and 60-70 cm. are strongly saline, at the depth of 70-120 cm. are medium saline soils. By type these soils have completely changed to chloride-sulfate type of salinization.

Due to stratification, a sharp decrease of Cl, SO_4, Ca and Mg content in the soil profile of section No.2 compared to section No.1 is detected (Table 2.1.2).

Note: Section #1-control, deposit (laid in September 16, 1999) at the beginning of the experiment.

Section No.2 - experimental plot (laid on the first year after completion of the 3-year experiment on September 15, 2004). The data of Table 2.1.2 show that under the influence of development, irrigation and biointensive technology there are dramatic changes in the chemical composition of meadow alluvial saline fallow soils.

The soil profile of section No. 1 is characterized by low humus content, the value of which varies between 0.54 and 0.77 % by weight of soil.

The content of gross nitrogen in accordance with the distribution of humus varies from 0.02 in the 10-20 cm layer to 0.036 % in the 47-70 cm layer. In the content of gross phosphorus there is some layering in its distribution by depth, which is apparently explained by the distribution of humus.

By content of assimilable form of phosphorus upper 0-10 and 10-20 cm. layers, and by potassium 0-10 cm. layers of soil belong to the category of highly provided, and in lower horizons these values pass to the categories of medium and low provided (Table 2.1.3).

For three years of development and application of biointensive farming technology, the humus content in the upper layers 0-10 and 10-45 cm. increased 5.3 (4.09 %) and 2.7 (1.68 %) times, respectively (section No. 2). Corresponding to humus in these horizons the content of gross phosphorus and nitrogen increased noticeably, and in the content of potassium some decrease was recorded throughout the depth of the section. According to the content of mobile forms of phosphorus and potassium, especially in the upper horizons, they should be categorized as highly provided.

Table 2.1.1.

Mechanical and micro aggregate composition of meadow alluvial irrigated saline soils

cut number	Depth, cm.	Composition	Weight of fractions in %							Physical clay	Sum of true aggregates
			>0.25	0.25-0.1	0.1-0.05	0.05-0.01	0.01-0.005	0.005-0.001	<0.001		
1	2	3	4	5	6	7	8	9	10	11	12
	0-10	Micro aggregate	0,8	0,2	20,4	71,2	2,4	0,6	4,4		23,1
		Mechanical	1,0	0,1	22,2	48,2	18,0	4,5	6,0	28,5	
		Difference	-0,2	+0,1	-1,8	+23,0	-15,6	-3,9	-1,6		
	10-20	Micro aggregate	0,8	0,2	22,6	70,4	1,2	0,6	4,2		29,4
		Mechanical	0,8	0,1	19,5	54,2	10,1	6,7	8,6	25,4	

		Difference	0	+0,1	+3,1	+26,2	-8,9	-6,1	-4,4		
	20-47	Micro aggregate	0,1	0,1	7,5	89,0	0,2	oh,h	2,8		33,3
		Mechanical	0,1	0,1	11,3	55,7	9,7	14,1	9,0	32,8	
		Difference	0	0	-3,8	+33,3	-9,5	-13,8	-6,2		
	47-80	Micro aggregate	0,2	0,1	14,1	79,9	0,8	0,6	4,3		19,6
		Mechanical	0,1	0,1	11,2	63,3	8,6	7,2	9,5	25,3	
1 16.09 1999		Difference	+0,1	0	+2,9	+16,6	-7,8	-6,6	-4,2		
	80-120	Micro aggregate	0,1	0,1	16,2	78,1	2,3	0,8	2,4		22,2
		Mechanical	0,2	0,1	16,6	55,9	8,0	12,8	6,4	27,2	
		Difference	-0,1	0	-0,4	+22,2	-5,7	-12,0	-4,0		
	0-10	Micro aggregate	30,0	1,0	26,7	34,5	6,4	1,0	0,4		17,2
		Mechanical	18,0	2,0	21,5	38,7	9,6	5,1	5,5	20,2	
		Difference	+12,0	-1,0	+5,2	-4,2	-3,2	-4,1	-5,1		
2 15.09. 2004	10-45	Micro aggregate	9,0	0,5	18,1	54,6	8,2	6,2	3,4		11,1
		Mechanical	1,8	0,2	17,4	51,7	9,1	10,2	9,6	28,9	
		Difference	+7,2	+0,3	+0,7	+2,9	-0,9	-4,0	-6,2		
	45-60	Micro aggregate	3,0	0,5	21,1	66,9	3,7	0,4	4,4		11,3
		Mechanical	1,5	0,1	18,6	60,0	8,0	4,0	7,8	19,8	
		Difference	+1,5	+0,4	+2,5	+6,9	-4,3	-3,6	-3,4		
	60-70	Micro aggregate	0,4	0,1	15,5	69,3	9,7	1,9	3,1		9,8
		Mechanical	0,1	0,1	14,3	63,1	7,6	6,1	8,7	22,4	
		Difference	+0,3	0	+1,2	+6,2	+2,1	-5,2	-5,6		
	70-120	Micro aggregate	0,1	0,1	11,6	72,1	8,3	3,4	4,4		30,1
		Mechanical	0,2	0,1	34,8	42,0	9,3	5,5	8,1	22,9	
		Difference	-0,1	0	-23,2	+30,1	-1,0	-2,1	-3,7		

Table 2.1.2.

Content of water-soluble salts in meadow alluvial irrigated saline soils.

	Depth, cm.	Dense residue, %	% mg-eq					Anion / cation	Na mg/eq in	By difference, %	Σ salts, %
			HCO_3	Cl	SO_4	Ca	Mg				
	0-10	6,50	0,027 0,44	1,907 53,78	1,683 35,00	0,205 10,23	0,229 18,82	89,22 29,05	60,17	1,38	5,42

1 16.09 1999	10-20	3,40	0,027 0,44	1,053 29,69	0,814 16,93	0,200 9,98	0,049 4,03	47,06 14,01	33,05	0,76	2,89
	20-47	3,10	0,024 0,39	0,770 21,71	0,971 20,20	0,170 8,48	0,070 5,75	42,30 14,23	28,07	0,64	2,63
	47-80	2,85	0,021 0,34	0,710 20,02	0,860 17,89	0,105 5,24	0,09 7,48	38,25 12,72	25,53	0,58	2,37
	80-120	3,31	0,027 0,44	0,714 20,13	1,114 23,17	0,200 9,98	0,085 6,99	43,74 16,97	26,77	0,61	2,74
2 15.09 2004	0-10	0,24	0,049 0,80	0,024 0,68	0,099 2,06	0,040 2,00	0,012 0,99	3,54 2,99	0,55	0,01	0,21
	10-45	0,27	0,030 0,51	0,014 0,39	0,144 2,99	0,035 1,75	0,015 1,23	3,89 2,98	0,91	0,02	0,24
	45-60	1,20	0,030 0,51	0,014 0,39	0,736 15,31	0,240 11,98	0,027 2,22	15,80 14,20	1,60	0,03	1,06
	60-70	0,71	0,021 0,34	0,024 0,99	0,428 8,90	0,130 6,49	0,021 1,73	10,23 8,22	2,01	0,04	0,66
	70-120	0,30	0,024 0,39	0,035 0,99	0,143 2,97	0,035 1,75	0,012 0,99	4,35 2,74	1,61	0,03	0,27

Leaching of CO_2 carbonates and SO_4 gypsum lower horizons was recorded. In the soil profile of section No.1 the content of carbonates in the uppermost horizon amounted to 9.07 %, which naturally killed with depth and in the layer 80-120 cm. amounted to 4.0 %. The SO_4 content of gypsum varies from 2.022 % in the 0-10 cm layer, and in the range of 1.012-1.399 % in the lower horizons.

Table 2.1.3.

Chemical composition of meadow alluvial irrigated saline soils.

Section	Depth, cm.	Humus %	NO3 %	P2O5 %	K2O %	P2O5 mg/kg	K2O mg/kg	CO2 %	SO4 Gypsum %
P-1 16.09 1999	0-10	0,77	00,33	0,227	2,34	137,3	315	9,07	2,022
	10-20	0,56	0,020	0,243	2,0	94,65	200	7,22	1,012
	20-47	0,68	0,027	0,270	2,06	26,73	160	5,9	1,399
	47-80	0,84	0,036	0,250	1,8	32,53	176	4,88	1,163
	80-120	0,54	0,034	0,230	1,37	14,93	190	4,00	1,273
P-2 15.09 2004	0-10	4,09	0,15	0,303	0,95	171,3	905	7,90	0,533
	10-45	1,68	0,07	0,343	1,2	115,3	320	8,44	0,599
	45-60	0,65	0,027	0,171	1,37	22,46	295	8,80	2,000
	60-70	0,39	0,039	0,189	1,25	20,93	260	9,19	1,181

	70-120	0,48	0,020	0,159	1,25	49,53	236	9,59	0,420

Carbonate regime in soil profile of section No.2 has a bright tendency of CO_2 value increase from 7.90 % in 0-10 cm layer to 9.19-9.60 % in lower layers. The maximum gypsum content was also recorded at the depth of 45-70 cm, i.e. in the lower horizons the formation of carbonate-illuvial horizon has started due to leaching of some elements from the upper soil layers.

2.2 Methodology of field and laboratory studies

In 1999-2004 on an open area of about 0.5 hectares a small plot field experiment was established according to the scheme given in Table 2.2.4.

Repetition of variants of field experiments is fourfold (Table 2.2.4). All repetitions are arranged in one tier. The total area of the field experiment is 980 m^2. The size of each plot - 24 m^2, the accounting area - 12 m^2. The distance between the first variant (Borozdovaya irrigation) and other variants is 3 meters. The other variants are protected from each other by a 60cm wide path. Each variant of experiments are placed in the form of beds of 1.5 X 16 meters.

During three years field experiments were set at the same place. We did not use various pesticides and mineral fertilizers. We used: - for irrigation household water meter, and for feeding two hundred liter plastic barrels. Vegetable seedlings were grown in boxes (greenhouse). In spring, the seedlings were planted in beds (April) following the method of John Jevons (1993). Each variation is divided into even two parts. A plot consists of 4 rows. On each accounting plot 80 pieces of seedlings - tomato (Volgogradsky-5/95), eggplant (Aurora) in staggered order (between plants distance - 35 cm) were planted. Tomato Volgogradsky-5/95 is zoned in all regions of the country. Medium-late - from sprouting to the beginning of ripening takes 120-140 days. Fruits are flat-rounded and rounded, smooth, weighing 80-100 g. The taste of fruits is mediocre (Buriev and Abdullaev, 1986).

Eggplant variety Aurora table and canning purpose, heat-resistant. Fruits are dark purple, weighing 200-250 g. Yield 700800 c/ha. It is desirable to plant it after root crops, onions, cabbage and melons. Eggplant succeeds better on light fertile soils, provided with moisture. Harvest the fruit when it reaches consumer maturity. The flesh is tender and tasty (Buriev and Abdullaev, 1986).

Tomato and eggplant contains fiber, which favorably affects the intestinal microflora, improve digestion. It helps to eliminate cholesterol from the human body.

Tomato and eggplant were harvested regularly at the onset of technological ripeness, avoiding overripening and coarsening of products. The yield was counted by a continuous method, weighing vegetables from the entire accounting plot, mathematical processing of yield was carried out in accordance with the methodology and recommendations of Peregudov V.N.

Soil samples were collected using soil drill (0-30cm), envelope method, in five-fold repetition - in spring, summer, fall and winter according to the experiment variants. Water samples were taken by months of the year from water supply networks of Nukus city and Kyzketken canal in accordance with the methodology of mobile laboratory HACH and LaMotte (USA).

Soil chemical analyses were determined (NH_4, NO_3, NO_2, P_2O_5, K_2O_5, pH, CL, SO_4, Ca, Mg, Mn, Cu, Fe) according to the requirement and recommendations of HACH mobile laboratory, LaMotte (USA).

Chemical analyses of soil and water were carried out in the Laboratory of Soil Science and Biotechnology of the Institute of Bioecology of the Karakalpak Branch of the Academy of Sciences of RUz under the supervision of B. Jolybekov.

Dynamics of water-soluble salts content in soil was studied by seasons of the year. Soil samples were collected by envelope method in fivefold repetition with soil drill (30 cm depth). Their contents were determined according to the methodology of UzNIIH (total salts, HCO_3, Cl, SO_4, Ca, Mg, Na, K).

Water accounting at each irrigation was carried out by household water meter The content of gross humus by seasons of the year (winter, spring, summer, and fall) was determined according to Tyurin.

Mechanical and micro aggregate composition were determined by the method of N.A. Kachinsky (1958).

Content of NH_4, NO_3, NO_2, P_2O_5, K_2O_5, pH, CL, SO_4, Ca, Mg, Mn, Cu, Fe by seasons of the year was determined in accordance with the requirements and recommendations of the mobile laboratory of HACH and LaMotte (USA).

Drinking and irrigation water quality was determined by months of the year according to the methodology of UzNIIH.

In 1999, organic fertilizer (compost) from calculation of 5 t/ha was applied in autumn under the grain field. Every year in October-February months winter grain and legume crops were grown (Table 2.2.4; Table 2.2.5). In March, using a mini tractor, siderates were chopped (3 t/ha) and plowed into the soil to a depth of 15-20 cm (Fig. 4). On the surface of the beds scattered compost 3-5 cm thick at the rate of 5 t/ha. and thoroughly mixed with a garden fork to a depth of 30 cm. In April, ground in the machine, dry alfalfa (2 t/ha) covered the surface of the beds (4-5 cm) and mulched, as well as planted vegetable seedlings according to the scheme of field experiments in a staggered order on each plot of 80 pieces

Table 2.2.4.

FIELD EXPERIMENT DESIGN

OPTIONS	Irrigation method	Norm, m³/ha		Organic fertilizer						Types of vegetable crops
				Dry, t/ha				Liquid, l/ha		
		Irrigation	Irrigation	Podzyab	Siderate	Before forming the bed	Pre-planting mulch	Korovniak	chicken droppings	
1	Borozdovaya (control)	400	6000	5	3	5	2	200	100	Tomato Eggplant
2	Rain	200	3000	5	3	5	2	200	100	Tomato Eggplant
3	Drip	100	1500	5	3	5	2	200	100	Tomato Eggplant
4	INTRODUCTION: Hose spraying	100	1500	5	3	5	2	200	100	Tomato Eggplant
5	INTRODUCTION: Plastic bottle	100	1500	5	3	5	2	200	100	Tomato Eggplant

Phenological observations (Fig. 8) and plant care (weeding, watering, etc.) were carried out from April. Liquid fertilizer diluted in June with cowdung (200 l/ha) and in July with chicken fertilizer (100 l/ha) was applied. In June-October, the harvested yield was recorded.

Since spring (April), irrigations were carried out, taking into account the needs of plants, according to the accepted norms and variants of the field experiment. Irrigation rate was supplied through a water meter. In the first variant full irrigation norm was supplied in one reception, and in other variants - in two receptions (in the morning-50% and in the evening 50%). At the end of October, after harvesting the plant stalks, organic fertilizers (compost) were applied at the rate of 5 t/ha and the soil was plowed with a mini tractor to a depth of 15-20 cm. Immediately after soil treatment we sowed seeds of grain and legume crops and further cultivated them.

The study was supported by the CAFE-Nukus organization under the auspices of the Swedish International Development Agency (SIDA).

The biointensive methodology uses specific techniques that are very different from other farming methods commonly practiced in developing countries. Deeply dug, high beds of loose, compost-enriched soil are the ideal place to grow the healthiest and most productive plants possible.

The width of the beds in the biointensive method is 1.5 meters. Plants are planted quite close to each other in order to get a high yield. Artificially created thick layer of soil made of compost and straw, plant leaves, which as a mulch will retain moisture and prevent the spread of weeds.

To obtain high yield with economical water and energy consumption using biointensive method, on irrigated saline soils, it is necessary to create high beds composed of loose soil, previously fertilized with compost. This biointensive bed is an optimal environment for the plant. Under such conditions, an optimal water-air regime in the soil is created. Plants develop a deeper root system and become stronger and more productive.

Biointensive farming method promotes the development of beneficial microflora around the root system of the plant in the soil. It creates good conditions for intensive reproduction of living organisms and earthworms. It promotes air penetration into the soil while fertilizing it, so that soil productivity increases dramatically in biointensive beds (John Jevons, 1993,1999).

Table 2.2.5.

Timing and types of agro-technical measures for 2000-2002

Months	*Agronomic measures*
January	Cultivation of winter cereals and legumes.
February	Cultivation of winter cereals and legumes.
March	Plowing of green winter cereals and legumes with a mini tractor to a depth of 15-20 cm (3 t/ha). Compost application (5 t/ha). Bed formation and soil tillage. Covering the beds with shredded dry alfalfa (mulch - 2 t/ha).
April	Planting from seedlings of tomatoes, peppers, eggplants and cabbage. Phenological observations. Care of plants. Watering.
May	Plant care. Watering.
June	Plant care. Watering. First fertilization with liquid cowpea (200l/ha). Harvesting and accounting of the crop.
July	Plant care. Watering. Second fertilization with liquid chicken manure (100 l/ha). Harvesting and accounting of the crop.
August September	Plant Care. Watering. Harvesting and record keeping.
October November	Harvesting and accounting of the crop. Application of organic fertilizers (compost) at the rate of 5 t/ha. Plowing the soil with a mini tractor to a depth of 1520 cm. Sowing winter cereals and legumes. Cultivation of winter cereals and legumes.

December	Cultivation of winter cereals and legumes

The growth of microflora in the soil is ensured by annual use of green siderate, compost and alfalfa straw (as mulch and organic fertilizer). Soil enrichment occurs due to the natural decomposition of plant stems and roots, as well as animal residues mixed with soil. This microflora, in most cases, is destroyed due to excessive use of pesticides and mineral fertilizers, as well as washed out of the soil at high irrigation rates.

To study the influence of different irrigation methods on the fertility of irrigated saline soils on the background of organic matter by the method of John Jevans (1993), we conducted field experiments in 1999, 2000 - 2002 according to the scheme of experiments (Table 2.2.4).

In the fall of 1999, each variant was placed on the experimental plot according to the adopted experiment scheme (Table 2.2.4). Then a full double digging was carried out once for rapid improvement of soil structure quality according to the method of John Jevons (1993) and John Jevons, Carol Cox (1999).

Double digging is the process of preparing the soil by digging it to a depth of two bayonets (about 60 cm). First of all, it is necessary to mark out a bed with a width of 1.5 meters and a length of -16 meters. To conduct a double digging remove soil from the trench, which is dug across, at the edge of the bed (depth and width of the trench-30 cm). Operation "Double digging" is carried out in the following order (Fig. 1):

1. Moisten the soil, loosen it and remove weeds. Then dig the bed with a pitchfork to a depth of 30 cm;
2. Apply a 4-5 cm layer of compost over the entire area to be dug;
3. thoroughly mix the compost to a depth of 30 cm;
4. Remove topsoil from trenches and carry it to the far end of beds to make compost and for seedlings;
5. Loosen the soil by another 30 cm;
6. Apply a 4-5 cm thick layer of compost to the loosened soil in the first trench;
7. Thoroughly mix the compost at the bottom of the first trench to a depth of 30 cm;
8. Excavate, the top of the second trench, carrying it to the top of the first trench;
9. Loosen the bottom of the second trench;
10. Apply a 4-5 cm thick layer of compost to the loosened soil of the lower part of the second trench;
11. thoroughly mix the compost at the bottom of the second trench to a depth of 30 cm;
12. Continue this process of double digging allowing good soil structure (repeat steps 8-11 in the remaining trenches);
13. Give the bed the desired shape. Then apply a layer of compost, straw, etc. of the required fertilizer evenly over the entire surface, burying it with forks to a depth of

10-15 cm. The bed is ready.

Composting is a biothermal process of mineralization and humification of organic matter, which takes place under aerobic conditions under the influence of microorganisms, mainly heat-loving. During composting, organic waste is heated to a temperature of 60 C, which has a devastating effect on fly larvae and pupae. Helminth eggs and pathogenic non-spore-forming microorganisms (Vasiliev and Filippova, 1988).

Laying the compost heap. When laying the compost heap, the materials are laid in layers (Fig. 2).

1. Loosen the soil with a garden fork to a depth of 30 cm in a 1.5 x 1.5 m area so that the heap has sufficient mass for the decomposition process to proceed normally. Loosening the soil will provide drainage and aeration;
2. Lay down a layer of coarse materials 8 cm thick that will promote air infiltration into the heap: twigs, small branches, corn or sunflower stalks. Reed or rose branches, etc.;
3. Lay layers of the compost sandwich, watering each layer as it is laid:
   - A 5 centimeter layer of dry materials;
   - A 5 centimeter layer of green materials;
   - A layer of soil that would lightly cover the lower layers.
4. Continue laying layers until the pile is 1.5 m high. Make sure that the top of the heap does not slip;
5. Cover the surface of the heap with an additional layer of soil to retain moisture inside;
6. Water the heap to make it moist. Check how moist the heap is in the middle; sometimes you may inadvertently over-moisten or over-dry the heap;
7. After 3-6 weeks, turn the heap over. The purpose of turning is to move dry, less decomposed materials to the inside of the heap and the most decomposed materials to the outside. Mix the materials from the old heap onto the new heap, putting the dry materials inside the heap. If necessary, water the new pile with water to ensure adequate hydration;
8. Allow the pile to "burn out" for 3-6 months. Compost is considered ready when:
   - most of the source material put in the pile becomes unrecognizable;
   - a fresh woody smell emanates from the pile, like meltwater in the spring;
   - the contents of the heap are dark brown in color, soft and crumbly (Fig.3).
9. If you do not use the compost when it is prepared, you should spread it out and dry it. It is very important not to let the heap decompose for too long.

Impregnating hose and its laying. Impregnating hose is made from old car tires in the United States and England. It is widely used in cottage plots and farms. These hoses are available in different diameters (d-15mm, d-20mm). They have thin holes (d-

0,3mm). Impregnating hoses are laid on the surface of biointensive beds at a depth of 13-15 cm and mulched with straw on top (Fig.5,6).

Plastic bottle and its tab. Plastic bottles with volume

1.5 liters are collected and its side side is made with a nail to make holes with a diameter of 1.5-2.0 mm. They are installed on the surface of biointensive beds upside down at 15-20 cm depth and mulched with straw on top (Fig. 7).

Chapter 3. Current chemical state of drinking, irrigation water in conditions of Nukus city

3.1 Changes in the chemical composition of drinking water

At present, the coverage of the population in cities with centralized water supply reaches 75% and in rural areas 40-50%. Uzbekistan adopted high norms of specific water consumption of 600-720 l/day for urban population and 150-180 l/day for rural population until 1995-2010 (Razakov 2001).

The population of the Republic of Karakalpakstan uses drinking water from water supply networks, supplied through the Tuyamuyun-Nukus-Takhtakupyr pipeline, tap water taken from irrigation networks, water from open ponds, wells, water from ECOS-50 desalination plants, and water taken from underground wells. Centralized water supply is provided to about 59.3% of the RK population, including 78.5% in cities and 44.2% in rural areas (Konstantinova et al. 2001).

Irrigation water supply to the territory of the Republic of Karakalpakstan fully depends on the conditions of formation and use of the transboundary river flow in the upper reaches. During the years of low water (2000-2001) water availability in these territories decreased to 40-50% compared to average annual values (Mironenkov, 2001).

At present, the coverage of population in cities with centralized water supply reaches 75%, and in rural areas 40-50%.

Uzbekistan adopted high norms of specific water consumption of 600-720 liters/day for urban population and 150-180 liters/day for rural population until 1995-2010 (Razakov 2001).

In some individual plots in Nukus, most farmers use tap water for irrigation and irrigation water from the Amudarya River.

Therefore, annual, seasonal and monthly monitoring of changes in water-soluble salts in the composition of drinking water of Nukus city and irrigation water of Kyzketken canal is very relevant.

Unreasonable management and use of water and land resources, development of unfavorable lands, violation of crop rotation, pollution with agrochemicals, waterlogging and salinization of lands have become the main cause of soil fertility decline and crop yields in the lower reaches of the Amudarya delta (Zhollybekov 1991,1995,1997, 2000, Zhollybekov 1996).

The object of the study is drinking water of Nukus city and irrigation water of Kyzketken canal.

Discharge of large volumes of polluted collector-drainage water directly into the riverbed without treatment worsened the quality of irrigation and drinking water in the lower reaches of the Amudarya River, especially in years of low water. Intensive soil salinization led to high river water consumption for land leaching. In the lower reaches of the Amudarya River, 2-3 times more irrigation water is used per unit of crop yield (Jolybekov 1995, 2000, Razakov 2001).

The weighted average content of hydrogen carbonate in the composition of drinking water of the city Nukus for three years remained almost constant (0.16 g/l). In the spring of 2001 the chlorine content reached a maximum (0.31 g/l) and in the summer of 2002 a minimum (0.15 g/l). In the lowest water year (2000), winter and spring chlorine content was lower (0.19 g/L) than the other seasons of the year (Table 3.1.1.). The weighted average chlorine content for the observed years ranged from 0.20-0.26 g/l.

The maximum sulfate content in drinking water was observed in spring 2001, and amounted to 0.44 g/l, and the minimum in spring 2002 (0.28 g/l). This is due to fluctuations in the Amu Darya river flow. The weighted average sulphate content depending on the volume of river flow varies within the range of 0.289-0.43 g/l (Table 3.1.1.).

Calcium content in the spring of 2000 in the drinking water of Nukus city, reached to the maximum (0.10 g/l), and in the fall of 2001 the minimum (0.07 g/l). The weighted average of calcium for the studied years remained almost unchanged.

Magnesium in drinking water reached to the maximum in spring 2002 and amounted to 0.07 g/l and minimum in summer 2002 (0.04 g/l). The weighted average magnesium content remained at the same level over the years of the study, i.e. 0.06 g/l -0.067 g/l (Table 3.1.1.).

The maximum sodium and potassium content reached 0.24 g/l in the summer of 2001 and the minimum in the spring of 2002 (0.13 g/l). The weighted average content of sodium and potassium over the years of the study ranged from 0.135 g/l to 0.21 g/l.

The total maximum mineralization of drinking water in winter 2002 reached up to 1.75 g/l, and the minimum in summer 2002 (0.73 g/l).

The weighted average mineralization content of drinking water in the studied years ranged from 1.1135 g/l (2002) to 1.253 g/l (2001).

Table 3.1.1.

Seasonal and annual changes in the content of water-soluble salts in drinking water of Nukus city, g/l / mg-eq

Years	Seasons	NCOz	Cl	SO4	Ca	Mg	Na+K	Amount of salts, g/l
	Winter	0,178	0,190	0,373	0,086	0,060	0,150	1,058
		2,92	5,5	7,7	4,32	5,56	6,30	
	Spring	0,16	0,189	0,442	0,100	0,058	0,186	1,130
		2,62	5,33	9,21	4,98	4,78	7,43	
	Summer	0,15	0,206	0,320	0,084	0,055	0,150	0,981
2000		2,52	5,8	6,7	4,2	4,6	6,3	
	Autumn	0,175	0,220	0,330	0,083	0,070	0,151	1,032
		2,86	6,2	6,93	4,16	5,8	6,03	
Average for the year.		0,165	0,201	0,366	0,088	0,061	0,159	1,050
	Winter	0,19	0,26	0,44	0,11	0,06	0,22	1,303
		3,26	7,30	9,30	5,55	5,40	8,91	
	Spring	0,18	0,31	0,48	0,139	0,07	0,21	1,412
		3,04	8,80	10,1	6,95	6,50	8,51	
	Summer	0,13	0,23	0,43	0,04	0,06	0,24	1,169
2001		2,28	6,60	9,15	2,25	5,65	10,13	
	Autumn	0,16	0,25	0,37	0,07	0,08	0,17	1,129
		2,65	7,26	7,83	3,53	7,13	7,08	
Average for the year.		0,167	0,26	0,43	0,090	0,067	0,21	1,253

2002	Winter	0,143	0,241	0,312	0,098	0,051	0,162	1,750
		2,34	6,80	6,50	4,90	4,25	6,49	
	Spring	0,17	0,234	0,284	0,082	0,072	0,131	0,970
		2,78	6,60	5,93	4,13	5,93	5,25	
	Summer	0,141	0,149	0,23	0,062	0,048	0,105	0,736
		2,32	4,20	4,80	3,10	4,00	4,22	
	Autumn	0,15	0,222	0,329	0,081	0,071	0,142	0,998
		2,49	6,26	6,86	4,06	5,86	5,69	
Average for the year.		0,15	0,21	0,289	0,081	0,060	0,135	1,113

Thus, in the epicenter of ecological crisis in the low-water years 20002001 drinking water quality deteriorated greatly and had a noticeable impact on soil salinization of experimental fields

3.2 Changes in the chemical composition of irrigation water

Irrigation water supply for the territory of the Republic of Karakalpakstan fully depends on the conditions of formation and use of the transboundary river flow in the upper reaches. During the low-water years of 2000-2001 water availability in these territories decreased to 40-50% compared to average annual values (Mironenkov, 2001).

Unreasonable management, use of water and land resources, development of unfavorable lands, violation of crop rotation, contamination of soils with agrochemicals, their waterlogging and salinization have caused the decline in soil fertility and yields of vegetable crops. Discharge of large volumes of polluted collector-drainage water has worsened the quality of irrigation and drinking water in the lower reaches of the Amu Darya River, especially in years of low water availability. Intensive soil salinization resulted in high river water consumption for land leaching. In the lower reaches of the Amudarya River, 2-3 times more irrigation water is used per unit of crop yield (Razakov, 2001).

The weighted average hydrogen carbonate content of irrigation water composition remained constant (0.16g/l) for the first two years and decreased to 0.13g/l in the third year.

In spring 2001 chlorine content reached a maximum (0.41 g/l) and in spring 2000 a minimum (0.18 g/l). The weighted average chlorine content for the observed years ranged from 0.204-0.296 g/l (Table 3.2.2.).

Table 3.2.2.

Seasonal and annual variation of water-soluble salts content in the Kyzketken canal irrigation water composition, g/l / mg-eq

Years	Seasons	HCO_3	Cl	SO_4	Ca	Mg	Na+K	Σ salts, g/l
	Winter	0,194	0,200	0,470	0,092	0,078	0,190	1,230
		3,10	5,80	9,79	4,60	6,43	7,83	

	Spring	0,160	0,189	0,420	0,09	0,065	0,174	1,110
		2,68	5,33	8,87	4,58	5,38	6,99	
2000	Summer	0,135	0,206	0,360	0,088	0,059	0,166	1,029
		2,21	5,80	7,7	4,40	4,86	6,64	
	Autumn	0,170	0,220	0,360	0,106	0,062	0,150	1,080
		2,82	6,2	7,6	5,3	5,13	6,22	
Average for the year.		0,165	0,204	0,402	0,094	0,066	0,170	1,112
	Winter	0,230	0,316	0,390	0,112	0,071	0,23	1,371
		3,84	8,90	8,30	5,60	5,85	9,59	
	Spring	0,160	0,410	0,450	0,140	0,069	0,28	1,520
		2,64	11,73	9,43	7,00	5,70	11,25	
	Summer	0,170	0,200	0,573	0,070	0,047	0,298	1,317
2001		2,08	5,66	11,90	3,53	3,90	12,24	
	Autumn	0,160	0,260	0,440	0,073	0,082	0,221	1,246
		2,64	7,33	9,33	3,66	6,80	8,84	
Average for the year.		0,18	0,296	0,463	0,099	0,067	0,257	1,363
	Winter	0,150	0,316	0,139	0,112	0,060	0,090	0,880
		2,60	8,90	2,90	5,60	5,05	3,75	
	Spring	0,170	0,170	0,340	0,088	0,062	0,130	0,973
		2,80	4,80	7,26	4,40	5,13	5,33	
	Summer	0,093	0,117	0,245	0,036	0,041	0,114	0,640
		2,60	3,30	5,10	1,80	3,40	4,75	
2002	Autumn	0,125	0,213	0,33	0,066	0,071	0,146	0,950
		2,05	6,00	7,00	3,34	5,86	5,85	
Average for the year.		0,134	0,204	0,263	0,075	0,058	0,12	0,861

The maximum sulfate content in irrigation water was observed in the summer of 2002, and was 0.58 g/l, and the minimum in the fall of 2000 (0.36 g/l). This was due to fluctuations in the flow of the Amu Darya River. The weighted average sulphate content, depending on the volume of river flow, varies within 0.20-0.29 g/l (Table 3.2.2.).

In irrigation water of Kyzketken canal, calcium content, in spring 2002, reached to the maximum and amounted to 0.14 g/l, and the minimum 0.09 g/l (spring 2000). The weighted average of sulfate per year, studied years ranged from 0.10 to 0.75 g/l.

Magnesium content in irrigation water, in spring 2001 reached to the maximum and amounted to 0.082 g/l, and the minimum was 0.041 g/l (summer 2002). The weighted average magnesium content during the years of study remained almost at the same level of 0.05- 0.06 g/l.

Sodium and potassium content in irrigation water of Kyzketken canal reached to the maximum and was 0.298 g/l (summer 2001), and the minimum 0.09 g/l (winter 2002). The weighted average for the year, studied years ranged from 0.11 to 0.25 g/l.

The total maximum mineralization of irrigation water reached 1.52 g/l in spring 2001 and minimum in summer 2002 (0.64 g/l). In recent years, as a result of increased river flow, their mineralization has decreased almost twice. Weighted average mineralization content of irrigation water of Kyzketken canal in the studied years ranged from 0.86 g/l (2002) to 1.36 g/l (2001).

Thus, in conditions of Nukus city in 2000-2001 low-water years the quality of drinking and irrigation water has deteriorated (Fig.9). Therefore, the source of salt accumulation in soil is not only groundwater, but also irrigation and tap water supplied for irrigation of agricultural crops in the conditions of Nukus city.

Chapter 4. Effect of different irrigation methods on water-soluble salt content of irrigated meadow alluvial saline soils

4.1 Change of water-soluble salts content under furrow irrigation method

Studies of water-soluble salts were carried out at special stationary sites. Observations were carried out from 2000 to 2002 by seasons of the year. The study was based on the method of sampling by continuous column recommended by V.A. Kovda (1947). Study of peculiarities of water-soluble salts regime on irrigated bog soils was carried out in many points of the Syrdarya delta. The study showed that salt movement is very slow and not in seasons, but in more multi-year long cycles, which are obviously related to the known cyclic changes of the named factors (Borovsky et al. 1959). Field experiments are the main supplier of experimental material on soil arogenesis and should fulfill the role of reference studies in solving the most complex genetic-geographical problems of agrogenesis (Kozlovsky, 2003). Heterogeneity of salt distribution along the profile expressed in three phases is fully confirmed. Methodology of soil salt regime study should provide separate characterization (Kozlovsky, Pankova, 1976).

The purpose of this chapter of the dissertation was to study the effect of different irrigation methods and different forms of irrigation under irrigation with mineralized tap water, on changes in the water-soluble salt content of irrigated meadow alluvial soils. The results of these studies are presented in Table 4.1.1.

The maximum content of hydrogen carbonate at furrow irrigation method was recorded in winter 2000 and 2001, and the maximum in fall 2002. It was found that the average annual value of HCO_3 gradually increases to the last year of the experiments and is: in 2000 - 0.088%; in 2001 - 0.085%; in 2002 - 0.091% (Table 4.1.1)

As it was already mentioned in the experimental plots vegetable crops were irrigated with mineralized tap water. Irrigation rate at furrow irrigation is 6000 m^3. Seven irrigations with the norm of 400 m^3 were made. Experiments were conducted against the background of biointensive farming technology. Changes of water-soluble salts content by seasons of 2000 had the following characteristic: in winter and autumn at salt sum 2,084 and 1,529%, Cl and SO_4 content 0,586 and 0,647 and 0,221 and 0,792% respectively, these soils belong to very strongly saline chloride-sulfate type of salinization. And in spring and summer with salt sum 0,791 and 0,590%, content of Cl and SO_4 0,248 and 0,228 and 0,248 and 0,082% respectively, these soils by degree belong to medium and strongly saline and sulfate-chloride type of salinization. These soils remained at the level of very strong and strong salinization by degree and chloride-sulfate salinization by type of salinization for all years of experiments. But still some decrease of salinization tension by the last 2002 year of experiments is revealed.

Thus, when using furrow irrigation method at irrigation rate ($6000m^{3)}$ on the background of biointensive technology in the arable layer of soil in the third year of research there is a slight decrease in the content of water-soluble salts, but in general they remain at the level of strong salinization.

Table 4.1.1.

Change of water-soluble salts content (%/ mg-eq) in 0-30 cm layer under furrow irrigation

Years	Seasons	NCOz	Cl	SO4	Ca	Mg	Na+K	Σ salts, %
	Winter	0,165	0,586	0,647	0,147	0,047	0,492	2,084
		2,7	16,5	13,475	7,35	3,92	21,405	
	Spring	0,055	0,248	0,228	0,060	0,024	0,176	0,791
2000		0,9	7,0	4,75	3,0	2,0	7,65	
	Summer	0,067	0,248	0,082	0,030	0,030	0,133	0,590
		1,1	7,0	1,7	1,5	2,5	5,8	
	Autumn	0,067	0,231	0,792	0,190	0,097	0,152	1,529
		1,1	6,5	16,5	9,5	8,0	6,6	
Annual average		0,088	0,328	0,437	0,106	0,049	0,238	1,248
	Winter	0,104	0,355	0,816	0,055	0,015	0,568	1,913
		1,7	10,0	17,0	2,75	1,25	24,7	
	Spring	0,090	0,078	0,240	0,032	0,023	0,119	0,582
2001		1,48	2,2	5,0	1,6	1,9	5,18	
	Summer	0,073	0,092	0,216	0,038	0,012	0,124	0,555
	Autumn	1,2	2,6	4,5	1,9	1,0	5,4	
		0,073	0,195	0,648	0,035	0,039	0,350	1,340
		1,2	5,5	13,5	1,75	3,25	15,2	
Annual average		0,085	0,180	0,.480	0,040	0,022	0,290	1,097
	Winter	0,054	0,078	0,168	0,046	0,023	0,054	0,423
		0,88	2,2	3,5	2,3	1,9	2,38	
	Spring	0,067	0,178	0,264	0,070	0,036	0,117	0,732
2002		1,1	5,0	5,5	3,5	3,0	5,1	
	Summer	0,116	0,319	1,104	0,115	0,069	0,515	2,238
		1,9	9,0	23,0	5,75	5,75	22,4	
	Autumn	0,128	0,142	0,096	0,095	0,024	0,031	0,516
		2,1	4,0	2,0	4,75	2,0	1,35	
Annual average		0,091	0,179	0,408	0,081	0,038	0,1790,2	0,977

4.2 Change of water-soluble salts content under rain-fed irrigation method

At irrigation by sprinkling water consumption decreased twice. The irrigation rate was set at 3000 m^3/ha. At irrigation rate of 200 m^3 15 irrigations with highly mineralized tap water were carried out. In this case, rain sprinklers were installed at a height of 50-60 cm from the soil surface.

It is known that sprinkling irrigation under the accepted agrotechnical conditions is accompanied by surface compaction and soil crust formation. This in turn increases water consumption for surface runoff. But under biointensive farming technology, the presence of mulch from plant residues prevents these negative phenomena to a greater extent.

At sprinkling irrigation with lower irrigation rate changes in water-soluble salts content by seasons of the year has some peculiarities. During the first year under sprinkling irrigation these soils were highly saline and chloride-sulfate salinity type. During the year the value of total salts varied from 1.098 to 1.620% by weight, and the value of Cl and SO_4 from 0.107 to 0.284 and from 0.505 to 0.890% by weight, respectively (Table 4.2.2).

In 2001 and 2002 in these soils salinity type remains as in 2000 chloride-sulphate type. In spring of 2001 salinity of these soils decreases to slightly saline (salt sum 0,591%), and in the rest time of the year salinity of soils remains at the level of strongly saline (salt sum within 1,041-1,886%).

Table 4.2.2.

Change of water-soluble salts content (%/ mg-eq) in 0-30 cm layer under rainfed irrigation method

Years	Seasons	HCO3	Cl	SO4	Ca	Mg	Na+K	Esols, %
	Winter	0,061	0,231	0,828	0,140	0,055	0,305	1,620
		1,0	6,5	17,25	7,0	4,5	13,25	
	Spring	0,049	0,156	0,542	0,136	0,010	0,205	1,098
2000		0,8	4,4	11,3	6,8	0,8	8,9	
	Summer	0,079	0,284	0,504	0,190	0,079	0,087	1,223
		1,3	8,0	10,5	9,5	6,5	3,8	
	Autumn	0,104	0,107	0,890	0,020	0,030	0,454	1,605
		1,7	3,0	18,55	1,0	2,5	19,75	
Annual average		0,073	0,194	0,691	0,121	0,043	0,262	1,386
	Winter	0,079	0,373	0,636	0,105	0,082	0,300	1,579
		1,3	10,5	13,25	5,25	6,79	13,09	
	Spring	0,051	0,064	0,302	0,062	0,026	0,086	0,591
2001		0,84	18	6,3	3,1	2,1	3,74	
	Summer	0,067	0,178	0,480	0,125	0,039	0,152	1,041
		1,1	5,0	10,0	6,25	3,29	6,6	
	Autumn	0,067	0,497	0,720	0,169	0,073	0,364	1,886
		1,1	14,0	15,0	8,25	6,0	15,85	
Annual average		0,066	0,278	0,534	0,114	0,055	0,225	1,274
	Winter	0,075	0,082	0,182	0,038	0,022	0,083	0,482
		1,24	2,3	3,8	1,9	1,85	3,59	
	Spring	0,052	0,074	0,216	0,065	0,013	0,073	0,493
2002		0,86	2,1	4,5	3,25	1,05	3,16	
	Summer	0,116	0,160	0,648	0,160	0,030	0,216	1,33
		1,9	4,5	13,5	8,0	2,5	9,4	
	Autumn	0,063	0,050	0,096	0,026	0,012	0,049	0,296
		1,04	1,4	2,0	1,3	1,0	2,14	
Annual average		0,076	0,091	0,285	0,072	0,019	0,105	0,650

By 2002, highly saline soils are found only in summer time (1.330%), and during the rest of the year salinization drops to the degree of slightly to moderately saline soils (Table 4.2.2).

The total maximum amount of water soluble salts in winter 2001 reached up to 1.886%, and the minimum in fall 2002 (0.296%). The weighted average of the sum of water-soluble salts under rainfed irrigation in the studied years ranged from 0.650 (2002) to 1.386% (2000).

Thus, under the rain-fed irrigation method during the growing season, even at reduced irrigation rate by two times (3000 m^3/ha) on the background of biointensive technology, some soil desalinization occurs in the arable layer except for the summer season.

4.3 Change of water-soluble salts content under drip irrigation method

Drip method is one of the progressive water-saving methods of irrigation of vegetable and other crops. Under this method, vegetable crops were irrigated with an irrigation rate of 1500 m^3, and an irrigation rate of 100 m^3. Drip hoses were installed at a height of about 5 cm above the soil surface. The change of the influence of this method of irrigation on the content of water-soluble salts by seasons of the year can be judged from the data of Table 4.3.3. As shown by the data of this table for all years of research soil of experimental fields is also characterized by chloride-sulfate type of salinization. In winter and summer of 2000 and 2001 the sum of salts in 0-30 cm layer was 0,948-0,911%, and 0,595 and 0,841%, respectively, by weight and they should be referred to weak- and medium-saline type, and in the rest of the year to strongly saline type. In 2002 the sum of salts exceeds one and is referred to strongly saline type.

Table 4.3.3.

Change of water-soluble salts content (%/ mg-eq) in 0-30 cm layer under drip irrigation method

Years	Seasons	HCO3	Cl	SO4	Ca	Mg	Na+K	Esols, %
	Winter	0,060	0,151	0,442	0,112	0,027	0,152	0,948
		1,05	4,24	9,2	5,6	2,3	6,6	
	Spring	0,049	0,224	0,941	0,177	0,077	0,262	1,730
2000		0,72	6,3	19,6	8,82	6,38	11,4	
	Summer	0,067	0,167	0,408	0,128	0,043	0,098	0,911
		1,1	4,7	8,5	6,4	3,6	4,3	
	Autumn	0,043	0,443	0,792	0,185	0,060	0,356	1,879
		0,7	12,5	16,5	9,2	5,0	15,5	
Annual average		0,055	0,246	0,645	0,150	0,052	0,217	1,367
	Winter	0,046	0,121	0,245	0,088	0,018	0,077	0,595
		0,76	3,4	5,1	4,4	1,5	3,36	
	Spring	0,073	0,177	0,792	0,150	0,018	0,315	1,525
2001		1,2	5,0	16,5	7,5	1,5	13,7	
	Summer	0,085	0,177	0,336	0,180	0,042	0,021	0,841
		1,4	5,0	7,0	9,0	3,5	0,9	
	Autumn	0,073	0,462	0,528	0,090	0,036	0,407	1,596
		1,2	13,0	11,0	4,5	3,0	17,7	

Annual average		0,069	0,234	0,475	0,127	0,028	0,205	1,139
	Winter	0,085	0,391	0,600	0,150	0,036	0,331	1,593
		1,4	11,0	12,5	7,5	3,0	14,4	
	Spring	0,055	0,195	0,432	0,110	0,015	0,199	1,006
2002		0,9	5,5	9,0	5,5	1,29	8,69	
	Summer	0,067	0,195	0,588	0,155	0,030	0,198	1,233
		1,1	5,5	12,25	7,75	2,5	8,6	
	Autumn	0,067	0,231	0,516	0,085	0,033	0,261	1,193
		1,1	6,5	10,75	4,25	2,75	11,35	
Annual average		0,068	0,253	0,534	0,125	0,028	0,247	1,256

Thus, when using drip irrigation method during the growing season of vegetable crops in the arable layer of irrigated saline soils on the background of biointensive technology did not occur the expected salinization at irrigation rate of 1500m^3. Apparently it is explained by the fact that at drip irrigation method at the specified irrigation rate and under climatic conditions of Nukus city the conditions for their restoration to the soil surface are actually created

4.4 Change in water-soluble salts content under in-soil irrigation method

The in-soil method is also a progressive water-saving method of irrigation, allowing to supply water directly to the roots of vegetable and other crops. We have tested two in-soil irrigation methods for vegetable crops.

1. When using impregnation hoses

This method is derived from America. Tested jointly with the staff of NASN and La Motte (USA). The work was performed on a contractual basis. The method is highly effective and convenient for work. In this method, the rate of irrigation and irrigation is also adopted respectively 1500 and 100 m^3, i.e. four times less than in furrow irrigation.

For all years of research at in-soil irrigation with the use of soaking hoses the following features are found in the character of distribution of water-soluble salts:

a) in water extract the content of chlorine ion is less than sulphates and these soils are characterized by chloride-sulphate type of salinization.

b) according to the content of total salts in winter and autumn periods of 2000 in arable 0-30 cm. layer of soil by degree belongs to very strongly saline, and in spring and summer - to strongly saline. The content of total salts in these periods of the year was 2.091-2.725 and 1.719 and 1.527% by weight, respectively (Table 4.4.4).

c) for the periods of 2001, these soils were highly saline and the value of total salts ranged from 1.218-1.590% by weight.

Table 4.4.4.

Change of water-soluble salts content (%/ mg-eq) in 0-30 cm layer under in-soil **(soaking hose) irrigation**

Years	Seasons	NCOz	Cl	SO4	Ca	Mg	Na+K	Σ salts, %
2000	Winter	0,104 1,7	0,355 10,0	0,972 20,2	0,190 9,5	0,054 4,5	0,413 17,95	2,091
	Spring	0,079 1,4	0,248 7,0	0,876 18,25	0,200 10,0	0,073 6,0	0,243 10,55	1,719
	Summer	0,079 1,3	0,337 9,5	0,636 13,25	0,200 10,0	0,550 4,50	0,220 9,55	1,527
	Autumn	0,150 2,6	0,106 3,0	1,608 33,5	0,060 3,0	0,061 5,0	0,715 31,1	2,725
Annual average		0,103	0,261	1,023	0,162	0,184	0,397	2,015
2001	Winter	0,183 3,0	0,177 5,0	0,744 15,5	0,090 4,5	0,045 3,79	0,351 15,29	1,590
	Spring	0,061 1,0	0,142 4,0	0,636 13,25	0,035 1,75	0,039 3,29	0,305 13,25	1,218
	Summer	0,043 0,7	0,213 6,0	0,672 14,0	0,070 3,5	0,033 2,75	0,332 14,45	1,363
	Autumn	0,061 1,0	0,284 8,0	0,780 16,25	0,040 2,0	0,173 14,25	0,207 9,0	1,545
Annual average		0,087	0,204	0,708	0,058	0,072	0,298	1,429
2002	Winter	0,043 0,7	0,213 6,0	0,372 7,75	0,090 4,5	0,048 4,0	0,137 5,95	0,903
	Spring	0,073 1,2	0,355 10,0	0,144 30,0	0,210 10,5	0,061 5,0	0,591 25,7	0,239
	Summer	0,063 1,04	0,106 3,0	0,264 5,5	0,044 2,2	0,011 0,9	0,148 6,44	0,636
	Autumn	0,085 1,4	0,107 3,0	0,252 5,25	0,069 3,25	0,036 3,0	0,078 3,4	0,623
Annual average		0,066	0,195	0,258	0,102	0,039	0,238	0,600

1) in all periods of 2002 the decrease of salinization degree of 0-30 cm. layer of soils was fixed.

By degree of salinization during the year soils of experimental fields had the following characteristic - in winter the sum of salts is 0,901% - medium saline; in spring - sum of salts - 0,239% - non-saline; in summer - sum of salts - 0,636% - medium saline; in average for the year - sum of salts - 0,600% - slightly and medium saline. By type - sulfate-chloride.

Quite noticeable soil desalinization at irrigation with the use of soaking hoses at reduction of irrigation rate four times testify to the improvement of this method in comparison with the previous one. With a relatively high efficiency of the method of in-soil irrigation with the use of soaking hoses, its wide implementation in our conditions is limited. The fact is that impregnating hoses are made in the USA, and the organization of its production in our conditions requires large resources and money. In connection with the above, we have developed a new method of in-soil irrigation using plastic bottles. This method is considered better than all previous methods, including the American, highly effective, water-saving, has no analogues in the Republic of Uzbekistan and in the world. The main material for its production can be the most common waste household plastic bottles, the amount of which we have unlimited.

2. When using plastic bottles (household waste)

This method was first developed by us. It is non-material-intensive, water-saving, productive, environmentally friendly, export-oriented, competitive, patentable, and simple in design. It uses ordinary rubber hoses of domestic production in place of impregnating hoses. It has no analogues in Uzbekistan and in the world.

Some peculiarities of salt regime of soils of experimental fields for the years of research can be traced by the data of Table 4.4.5.

During 2000-2001 soils of experimental fields are classified by type as chloride-sulfate soils. During the mentioned period they are characterized as very strong and highly saline soils by salt sum.

The content of sum of salts in winter time is 3.472 - 3.631%, and in the rest of the year 0.890 - 1.701% to soil weight. In 2002 the character of salinization changes somewhat and passes to sulfate-chloride type. In winter and spring of the same year these soils remain in the category of strongly and very strongly saline by the sum of salts, and in summer and autumn they are moderately and strongly saline. The value of salt sum is: in winter and spring respectively 2.768 and 1.028%, and in summer and fall 0.581 and 0.665% by weight.

Table 4.4.5.

Change of water-soluble salts content (%/ mg-eq) in 0-30 cm layer under in-soil irrigation method (plastic bottle)

Years	Seasons	NCOz	Cl	SO4	Ca	Mg	Na + K	Σ salts, %
2000	Winter	0,122 2,0	0,781 22,0	1,560 32,5	0,180 9,0	0,289 24,0	0,540 23,5	3,472
	Spring	0,061 1,0	0,124 3,5	0,564 11,75	0,120 6,0	0,060 5,0	0,121 5,25	1,050
	Summer	0,091 1,5	0,391 11,0	0,672 14,0	0,130 6,5	0,049 4,0	0,368 16,0	1,701
	Autumn	0,110 1,8	0,266 7,5	0,679 14,15	0,090 4,5	0,049 4,0	0,344 14,95	1,538
Annual average		0,096	0,390	0,869	0,130	0,112	0,343	1,940
2001	Winter	0,122 2,0	0,355 10,0	1,992 41,5	0,200 10,0	0,042 3,5	0,920 40,0	3,631
	Spring	0,067 1,1	0,177 5,0	0,360 7,5	0,140 7,0	0,006 0,5	0,140 6,1	0,890
	Summer	0,055 0,9	0,213 6,0	0,648 13,5	0,075 3,75	0,036 3,0	0,314 13,65	1,341

	Autumn	0,049 0,8	0,142 4,0	0,780 16,25	0,060 3,0	0,021 1,75	0,375 16,3	1,427
Annual average		0,073	0,222	0,945	0,119	0,026	0,437	1,822
2002	Winter	0,110 1,8	0,781 22,0	0,426 12,0	1,344 28,0	0,200 10,0	0,049 4,0	2,768
	Spring	0,049 0,8	0,124 3,5	0,092 2,6	0,552 11,5	0,028 1,4	0,003 0,3	1,028
	Summer	0,049 0,8	0,391 11,0	0,099 2,8	0,245 5,1	0,026 1,3	0,008 0,7	0,581
	Autumn	0,049 0,8	,266 7,5	0,064 1,8	0,341 7,1	0,028 1,4	0,008 0,7	0,665
Annual average		0,064	0,390	0,170	0,620	0,070	0,017	1,260

As the data of Figure 10 shows the content of the sum of salts in the arable horizon is noticeably higher in comparison with other irrigation variants, especially to the variants where rain irrigation and soaking hoses were used. It seems to us that this can be explained in two ways.

1. The use of plastic bottles for irrigation creates conditions for salt migration from deep layers in the arable horizon.
2. In some degree of imperfection of this method that it is necessary to continue research in this direction. Nevertheless, it can be concluded that the in-soil irrigation method with the use of plastic bottles by its indication is a water-saving technology (irrigation rate of 1500 m^3/ha) under the biointensive method of farming, and can be recommended for a wide production test with further implementation in production in the cultivation of vegetable and melon crops.

Chapter 5. Influence of different irrigation methods and biointensive technology on fertility of irrigated meadow alluvial saline soils

5.1 Influence of furrow irrigation method on changes in the content of chemical elements in soil

When growing vegetable crops in the traditional furrow irrigation method we did not apply autumn-winter leaching irrigation on irrigated saline soils, i.e. in autumn the content of water-soluble salts in irrigation water varies from 0.95% (2002) to 1.25% (2001), which in turn will lead to secondary salinization of irrigated soils. Experimental plots were irrigated with tap water of high salinity, which in turn also induced secondary salinization of soils. At irrigation rate of 6000 m^3 and experimental fields were irrigated 7 times (April, May, June, July, August, September and October) with irrigation rate of 880 m^3. In autumn 1999, organic fertilizer (compost) was applied under the grain field at the rate of 5 t/ha and winter cereals (wheat) and leguminous crops (vetch) were grown. In winter in January soil samples were taken from the arable layer (030cm) for laboratory analyses.

The results of the study showed that the weighted average annual humus content was 1.81% in 2000, 2.247% in 2001 and 2.573% in 2002 (Table 5.1.1). This indicates that traditional furrow irrigation, annual application of organic fertilizers in the form of liquid manure, compost and siderate create conditions for humus accumulation in the arable horizon of soils.

The weighted average content of ammonium nitrogen in 2000 amounted to 65.5 mg/kg, and in the third year of the experiment decreased to 62.1 mg/kg. This is explained by the fact that part of ammonium nitrogen is intensively used by plants, and also washed with irrigation water into deep soil layers.

Increase of nitrate and nitrite nitrogen content was recorded. This is due to annual cultivation of grain and legume crops and application of organic matter their values increase respectively 10.9 mg/kg (2000) to 17.3 mg/kg (2000) and 37 (2000) to 50.0 mg/kg (2000) (Table 5.1.1) There is a gradual accumulation of mobile forms of phosphorus and potassium from year to year. The pH value remains almost at the same level of about 7.1-7.5. (Table 5.1.1). Irrigation with tap water with high mineralization has its influence on the content of chemical elements of soils for all years of research.

The maximum chlorine content was recorded in summer 2002 (8680 mg/kg) and the minimum in fall 2001 (1220 mg/kg). Its average annual value in 2002 almost doubled in comparison with 2000 (Table 5.1.2).

The maximum sulfate content was observed in spring 2002 and the minimum in winter 2000 (115.9 mg/kg). The weighted average of sulfate for the year, during the experiment gradually increased from 233 to 344.6 mg/kg.

Table 5.1.1.

Seasonal and annual changes in humus content, mobile forms of nitrogen, phosphorus, potassium and pH in 30 cm layer under furrow irrigation method

Years	Seasons	Humus %	N-NH4 mg/kg	N-NO2 mg/kg	N-NO3 mg/kg	P2O5 mg/kg	K2O mg/kg	pH
2 0 0 0	Winter	1,93	46,5	8,4	37	36,8	304,7	6,8
	Spring	1,97	92,0	14,8	50	16,1	398,3	8,1
	Summer	1,89	75,5	19,7	40	40,5	201,1	7,8
	Autumn	1,5	48,0	0,8	22	21,6	304,7	7,1
Annual average		1,82	65,5	10,9	37,25	28,7	302,2	7,5
2 0 0 1	Winter	2,302	54,3	8,1	44,0	42,1	401,1	7,0
	Spring	2,505	76,5	14,1	54,0	39,1	467,3	7,5
	Summer	2,100	64,3	21,2	41,0	40,4	253,1	6,9
	Autumn	2,083	51,1	7,2	37,0	24,1	308,1	7,2
Annual average		2,247	61,5	12,6	44,0	36,4	357,4	7,1
2 0 0 2	Winter	2,234	76,4	12,4	48,0	40,4	561,3	7,2
	Spring	2,640	66,4	16,4	62,0	37,1	520,3	7,1
	Summer	2,662	51,1	22,4	48,0	40,1	466,1	6,8
	Autumn	2,757	54,4	18,1	42,0	28,1	428,7	7,3
Annual average		2,573	62,1	17,3	50,0	36,42	494,1	7,1

In all years of the study, the maximum calcium content was recorded in spring. This seems to be related to the well-known position that solubility of carbonate salts, including CaCOs increases at low temperature of soil solution, i.e. in spring and winter. Maximum magnesium content was also observed in spring and ranged from 546.1-567.8 mg/kg. This is apparently due to the increase of magnesium content in irrigation and drinking water in winter and spring seasons.

Its minimum content was recorded in winter 2002 (146.0 mg/kg). In general, an increase in magnesium is detected towards the end of the research and it varied between 368.8 mg/kg in 2000 and 399.4 mg/kg in 2002.

Manganese content is more stable both during the year and during all years of research. But still some enrichment of soils with this element by the final year 2002 is traced. This testifies to the improvement of fertility of irrigated meadow alluvial soils under biointensive method of farming. The average annual value of manganese for the years of observation increased from 90.9 mg/kg to o 97.6 *mg/kg*.

Under the influence of biointensive technology enrichment of soils with copper trace element is detected. The distribution of copper by seasons of the year and by average indices for the years of research has its own peculiarities. Its minimum value for all years of experiments was recorded in summer time (Table 5.1.2)

The maximum value of iron is found in all years of observation in winter time, and the minimum value in autumn. In general, in the arable horizon of the studied soils in the last two years iron accumulated 6-7 times more than at the beginning of the research.

Thus, at application under vegetable crops, furrow irrigation in conditions of annual application of organic fertilizers in the rate of 15 t/ha in the form of compost, siderate and mulch there is an increase in humus in comparison with the initial indicator 1.42 times (Fig. 11).

During intensive cultivation of vegetable crops on the background of organic matter and furrow irrigation humus accumulates very slowly, and the content of ammonium nitrogen decreases.

Accumulation of nitrate and nitrite nitrogen, phosphorus (Fig.13) and potassium in soils is revealed, and soil pH remains constant. Chlorine content increases two times, sulfate and calcium content increases insignificantly. In the contents of magnesium, manganese and copper there was a noticeable increase, and iron increased 6-7 times.

Table 5.1.2.

Seasonal and annual changes of chemical substances content in 030 cm layer under furrow irrigation method

Years,	Seasons	CL mg/kg	SO4 mg/kg	Ca mg/kg	Mg mg/kg	Mn mg/kg	Cu mg/kg	Fe mg/kg
2000	Winter	2840	115,9	9329	198,4	99,7	2,2	0,6
	Spring	2520	184,0	11209	547,2	82,8	3,0	0,4
	Summer	1740	288,2	8227	422,4	99,7	1,8	0,2
	Autumn	1440	344,2	6128	307,2	99,7	2,4	0,8
Annual average		2135	233,0	8723	368,8	95,4	2,3	0,5
2001	Winter	2815	133,4	5720	166,0	86,5	3,1	5,1
	Spring	2560	351,0	10556	567,8	82,1	2,4	4,0
	Summer	1280	296,1	7698	465,7	96,2	2,2	4,4
	Autumn	1220	200,0	5158	398,1	98,9	2,8	1,8
Annual average		1968	245,1	7283	399,4	90,9	2,6	3,8
2002	Winter	2460	298,2	6420	146,0	98,9	4,1	4,8
	Spring	1268	484,1	10240	546,1	107,1	3,8	4,0
	Summer	8680	276,0	6200	465,7	82,1	2,2	2,2

	Autumn	7844	320,0	7460	424,4	98,9	2,1	0,8
Annual average		5063	344,6	7580	395,5	96,7	3,0	2,9

5.2 Influence of rain-fed irrigation method on changes in the content of chemical elements in soil

As it was already mentioned, the main purpose of our research is to enrich the arable 0-30 cm. layer of soil with organics and create fertile soil, and on it without the use of toxic chemicals and mineral fertilizers to obtain high, ecologically clean yield of vegetable crops; secondly, to select the optimal options for irrigation of vegetable crops with minimum irrigation water consumption. In this section we will analyze the behavior of chemical elements in soil under sprinkling irrigation. We will point out that sprinkling irrigation is associated with compaction of upper soil layers, formation of soil crust and consumption of part of water for surface runoff.

At initial condition these soils have low humus content. Its content in 0-30 cm layer is 0.67%, phosphorus and potassium content is high. Under bio-intensive technology and under sprinkling irrigation (watered with mineralized tap water) there were certain changes in the content of humus, nutrients and other chemical elements in the soil (Table 5.2.3). Under bio-intensive technology in the first years of the study the soil formation process is accompanied by rapid humus accumulation in the arable horizon of meadow alluvial saline soils and they passed to the category of highly humus-sufficient soils. Average annual humus content gradually increases from 1.66% in 2000 to 2.709% in 2002.

Table 5.2.3.

Seasonal and annual changes in humus content, mobile forms of nitrogen, phosphorus, potassium and pH in 30 cm layer under rainfed irrigation method

Years	Seasons	Humus %	N-NH4 mg/kg	N-NO2 mg/kg	N-NO3 mg/kg	P2O5 mg/kg	K2O mg/kg	pH
	Winter	1,44	13,0	4,7	148,0	23,6	159,0	6,4
	Spring	1,99	15,1	53,0	152,1	12,6	511,6	8,4
	Summer	1,66	27,5	6,9	58,0	29,9	521,0	7,5
2000	Autumn	1,55	33,5	5,8	152,1	20,7	468,3	7,6
Annual average		1,66	56,2	17,6	127,5	21,7	414,9	7,5
	Winter	3,193	4,5	3,0	152,1	28,8	139,9	6,8
	Spring	2,435	37,5	14,8	129,0	20,7	370,1	7,5
	Summer	2,196	27,5	5,8	152,1	15,2	511,6	7,8
2001	Autumn	2,320	37,5	4,7	152,1	15,2	520,1	7,0
Annual average		2,525	26,7	7,1	146,3	19,9	385,4	7,3
2002	Winter	2,890	115,0	4,2	54,0	16,2	152,0	6,4

	Spring	3,526	22,2	4,9	78,0	19,5	103,5	7,0
	Summer	2,723	29,6	5,6	70,0	23,1	511,6	7,4
	Autumn	2,698	36,9	5,7	130,8	18,2	472,3	7,2
Annual average		2,709	50,9	5,1	83,2	19,2	309,8	7,0

Maximum nitrate nitrogen content was recorded in the fall of 2000 and 2002 (152.1%), and in 2002 its value decreased almost 3 times. Average annual value of this element decreased by 1.5 times by 2002. Similar regularity is found in the content of mobile forms of phosphorus and potassium. The results of the research have shown that for all years of research these soils were in the state of low content of mobile forms of P2O5. As for K2O on the contrary, these soils were highly and highly supplied. The only reason for this phenomenon may be the fact that compost making materials contained insignificant amount of P2O5. This in turn requires that items with high phosphorus content should be added to the set of composting materials.

Table 5.2.4 provides data on the change in the content of chemical substances in the 0-30 cm layer under rainfed irrigation. Of these elements the content of trace elements Mn and Cu is of practical importance. According to the average annual value for the years of research, soils of experimental fields are normally provided with manganese, except for winter time, when they have insufficient amount of manganese.

Table 5.2.4.

Seasonal and annual changes in the content of chemical substances, in 0-30 cm layer under rainfed irrigation method

Years	Seasons	CL mg/kg	SO4 mg/kg	Ca mg/kg	Mg mg/kg	Mn mg/kg	Cu mg/kg	Fe mg/kg
2000	Winter	2200	121,8	1120	595,2	35,0	3,2	1,8
	Spring	2640	344,2	2960	465,6	100,1	0,3	0,2
	Summer	2780	344,9	608	290,4	99,7	0,2	2,0
	Autumn	1560	172,1	2832	326,4	100,7	2,4	0,8
Annual average		2295	245,7	1880	404,4	83,8	1,5	1,2
2001	Winter	1760	110,4	1296	278,4	42,4	15,2	1,7
	Spring	1680	344,2	8916	143,4	99,7	0,6	0,4
	Summer	840	156,9	3982	278,4	99,7	1,6	0,2
	Autumn	1040	344,2	1680	720,0	99,7	0,5	1,1
Annual average		1330	238,9	3968,5	355,0	85,3	4,4	0,8
2002	Winter	840	144,1	2200	440	38,2	1,2	2,6
	Spring	1100	156,6	4200	560	85,5	0,2	1,1
	Summer	400	172,1	1000	120	99,7	0,9	0,8
	Autumn	980	348,0	2300	630	98,9	1,8	0,9
Annual average		830	205,2	2425	43,7	80,5	1,0	1,3

The trace element Cu in these soils is deficient in spring and summer time. In the average annual cycle these soils are enriched with copper.

Thus, when growing vegetable crops under biointensive technologies with the use of rain-fed irrigation with annual irrigation rate of 3000 m2/ha there is an increase in humus content in comparison with the initial state of soils in 1.63 times (Fig. 11). The content of ammonium nitrogen-2 times, nitrate nitrogen-3 times, nitrite nitrogen-1.5 times, phosphorus-1.2 times, potassium-1.3 times, chlorine 1.5-2.0 times, sulfate-1.2 times decreases in the studied soils. The content of calcium-2.0 times. The content of magnesium, manganese, copper and iron increases markedly. It is revealed that copper and iron accumulate in the cold season.

5.3 Influence of drip irrigation method on changes in the content of chemical elements in the soil

Vegetable crops were grown on the background of organic fertilizers and irrigated with irrigation rate of 1500 m^3/ha using drip irrigation method. Watering was done 15 times with the rate of 100 m^3/ha.

The conducted research shows that at application of drip irrigation method for three years average weighted humus content in 0-30 cm layer varied from 1.39 to 2.788% by weight and these soils are referred to average humus-supplied soils (Table 5.3.5.). Ammonium nitrogen content increased almost 2 times in comparison with 2000 and in 2002 amounted to 49.5 mg/kg (Table 5.3.5).

The weighted average content of nitrite and nitrate nitrogen, phosphorus and potassium decreases markedly. The pH indicator is characterized by the value within 6.8-7.5%. Provision of these soils with mobile forms of phosphorus in 2000 and 2001 is low, and in 2002 - very low. Provision of soils of experimental fields with mobile forms of potassium in 2000 and 2002 is high, and in 2002 - low.

Thus, when using drip irrigation for vegetable crops on the background of biointensive technology in the arable layer of soil humus content compared to the initial state increases-1.7 times, ammonium nitrogen-2 times. The weighted average content of nitrate, nitrite nitrogen, phosphorus and potassium decreases.

Table 5.3.5.

Seasonal and annual changes in humus content, mobile forms of nitrogen, phosphorus, potassium and pH in 30 cm layer under drip irrigation methods

Years	Seasons	Humus %	N-NH4 mg/kg	N-NO2 mg/kg	N-NO3 mg/kg	P2O5 mg/kg	K2O mg/kg	pH
2000	Winter	1,39	44,5	3,0	152,1	28,8	139,9	6,8
	Spring	1,89	37,5	14,8	129,0	20,7	370,1	7,1
	Summer	1,60	27,5	5,8	152,1	15,2	511,6	7,6
	Autumn	1,91	37,5	4,7	152,1	15,2	520,1	6,8
Annual average		1,69	26,7	7,0	146,3	19,7	450,2	7,7
20	Winter	2,44	57,0	6,8	176	18,5	110,3	6,1
	Spring	2,787	31,0	11,3	102	33,4	146,1	7,4

	Summer	2,228	59,0	15,1	151	19,1	244,1	7,0
	Autumn	2,106	37,9	5,9	121	20,1	239,8	6,8
Annual average		2,392	46,2	7,2	137,5	22,7	185,0	6,8
2002	Winter	2,446	49,0	4,6	151,0	11,4	288,1	7,1
	Spring	2,192	46,0	6,8	133,0	3,8	255,8	8,1
	Summer	2,342	57,0	9,4	122,0	8,6	572,9	7,6
	Autumn	2,788	46,1	4,8	110,0	13,1	639,2	7,1
Annual average		2,442	49,5	6,8	129,0	9,2	439,0	7,5

Of the water-soluble salts. chlorine ions decreases 2 times. Copper decreases 1,5-2,0 times, and iron increases 2,0-2,5 times. Provision with microelements of these soils is ambiguous. Their provision with manganese is normal, and copper is increased (Table 5.3.6).

Table 5.3.6.

Seasonal and annual changes of chemical substances content (mg/kg) in 30 cm layer under drip irrigation method

Years	Seasons	CL mg/kg	SO4 mg/kg	Ca mg/kg	Mg mg/kg	Mn mg/kg	Cu mg/kg	Fe mg/kg
2000	Winter	1760	110,4	1296	278,4	42,2	15,2	1,7
	Spring	1680	344,2	8916	143,7	99,7	0,6	0,4
	Summer	840	156,9	3982	278,4	99,7	1,6	0,2
	Autumn	1040	344,2	1680	720,0	99,7	0,5	1,1
Annual average		1330	238,9	3968,5	678,4	85,3	4,4	0,8
2001	Winter	660	241,3	2330,0	502,1	21,1	7,8	7,1
	Spring	695	344,6	9586,0	629,8	87,7	1,4	1,1
	Summer	295	230,8	498,7	422,5	40,5	7,1	3,2
	Autumn	265	201,7	1965,2	410,0	93,7	13	1,8
Annual average		478,7	254,6	3594,9	491,1	60,7	7,3	3,0
2002	Winter	840	156,7	1200	278,4	57,1	8,1	8,4
	Spring	920	344,2	6600	620,0	99,7	1,6	0,8
	Summer	800	172,1	4860	464,7	76,1	1,6	1,1
	Autumn	600	344,2	2840	400,4	99,7	0,6	1,4
Annual average		790	254,3	3875	440,8	83,1	2,9	2,9

5.4 Influence of in-soil irrigation method on changes in the content of chemical elements in the soil

For irrigation of vegetable crops we used for the first time impregnating hoses made abroad (USA, England) designed for water supply into the soil.

At application of in-soil irrigation method with the help of soaking hoses the increase of humus content in 1.52 times is observed and they are characterized as medium humus-supplied soils.

Some increase of N-NO3 content and decrease of phosphorus value are traced. In general and in this variant these soils are not provided with mobile forms of phosphorus. Provision with potassium of soils of experimental fields is elevated and high. Deficiency of phosphorus in all probability is connected with small amount in composts, which is applied as organic fertilizers (Table 5.4.7.).

Chlorine, sulfate content gradually decreases, calcium-2.3 times and magnesium-1.3 times increases. Weighted average manganese content decreases from 108.7 mg/kg to 99.7 mg/kg, and these soils remain at the level of normal and high magnesium supply. Honey content ranges from 1.2-1.9 mg/kg and the soils are elevated in copper. Iron content remains at the same level (Table 5.4.8.).

Thus, using soaker hose for in-soil irrigation increases humus, nitrite nitrogen, potassium, calcium, magnesium, honey and decreases ammonium nitrogen, nitrate nitrogen, phosphorus and iron content remains at the same level.

Table 5.4.7.

Seasonal and annual changes in humus content, mobile forms of nitrogen, phosphorus, potassium and pH in 30 cm layer, under intra-pond irrigation method (soaking hose)

Years	Seasons	Humus %	N-NH4 mg/kg	N-NO2 mg/kg	N-NO3 mg/kg	P2O5 mg/kg	K2O mg/kg	pH
2000	Winter	1,25	89,0	18,5	144,0	31,0	207,0	7,2
	Spring	1,98	30,5	23,0	152,1	36,8	511,6	7,6
	Summer	1,69	21,0	29,2	129,0	28,8	511,6	8,0
	Autumn	1,84	28,5	13,1	132,0	13,5	288,6	7,8
Annual average		1,69	42,4	20,9	139,2	27,5	379,7	7,6
2001	Winter	2,084	44,5	13,0	152,1	28,8	139,9	6,8
	Spring	2,079	37,5	14,8	129	20,7	370,1	7,5
	Summer	1,873	27,5	5,8	152,1	15,2	511,6	7,8
	Autumn	2,070	37,5	4,7	152,1	15,2	520,1	7
Annual average		2,026	36,70	14,5	146,3	19,97	385,4	7,3
2002	Winter	2,605	28,1	15,4	134,7	20,1	176,0	8,1
	Spring	2,476	27,5	12,7	152,1	18,4	511,6	7,5
	Summer	2,507	27,5	16,4	152,1	12,5	520,0	7,0
	Autumn	2,741	31,4	14,1	152,1	14,4	516,4	7,1
Annual average		2,582	28,6	14,6	147,7	16,3	431,0	7,6

As a result of our search, we created a new water-saving irrigation system from household waste using plastic bottles.

Table 5.4.8.

Seasonal and annual changes in the content of chemical substances in the 0-30 cm layer under the in-line irrigation method (soaking hose)

Years	Seasons	CL mg/kg	SO4 mg/kg	Ca mg/kg	Mg mg/kg	Mn mg/kg	Cu mg/kg	Fe mg/kg
2000	Winter	1800	156,9	1680,0	100,8	135,8	2,0	1,1
	Spring	1560	344,2	2968,0	590,4	99,7	2,4	2,4
	Summer	1400	344,9	723,2	411,8	99,7	0,6	0,6
	Autumn	600	344,2	1560,0	250,0	99,7	0,2	0,8
Annual average		1340	297,5	1732,8	338,2	108,7	1,3	1,2
2001	Winter	1760	110,4	1296,0	278,4	42,4	5,2	1,6
	Spring	1680	344,2	8916,0	143,4	99,7	0,6	2,4
	Summer	840	156,9	3982,0	278,4	99,7	1,6	0,2
	Autumn	1040	344,2	1680,0	720,0	99,7	0,5	1,1
Annual average		1330	238,9	3968,5	355,0	85,3	1,9	1,3
2002	Winter	1120	160,1	1428	472,0	99,7	2,1	2,4
	Spring	1660	288,2	8400	486,4	99,7	1,4	1,6
	Summer	840	168,1	4400	386,0	99,7	0,6	0,6
	Autumn	1200	240,4	2220	472,0	99,7	0,8	0,1
Annual average		1205	214,2	454,1	454,1	99,7	1,2	1,1

In the new water-saving irrigation system, water is delivered inside the soil directly to the plant roots. In this method surface evaporation of irrigation water is completely excluded. The arable soil layer is not compacted due to irrigation. Good conditions are created for the reproduction of microflora. There is no need for repeated loosening of the top layer of soil during the vegetation period of plants.

When using plastic bottles for irrigation of vegetable crops as irrigation systems, there is an increase in humus content and soils for all years of the study average ammonium and nitrate form of nitrogen are accumulated mainly in winter-spring seasons of the year. In summer-autumn vegetation periods (summer 2000) their amount decreases to 14 mg/kg and 22 mg/kg, respectively (Table 5.4.9.). The weighted average nitrite content gradually decreases with each year. Soil pH indicator is normal.

For all years of research and in this variant of the experiment low provided with mobile forms of phosphorus. In connection with further either it is necessary to apply phosphorus fertilizers or to include materials containing phosphorus for compost making. During the years of research the soils of experimental fields are moderately and elevatedly provided with assimilable forms of potassium (Fig.14). Some decrease in manganese content was recorded in the last years of the study (Table 5.4.10.) in the first two years (2000, 2001) soils are normally supplied, and in 2002 they are insufficiently supplied with manganese. Excessive accumulation of copper is detected and according to provision with this trace element it is characterized as enrichment province (RKK soils, Khojaili district, book 1,

Tashkent - 1995).

Table 5.4.9.

Seasonal and annual changes in humus content, mobile forms of nitrogen, phosphorus, potassium and pH in the 0-30 cm layer under the intra-pond (plastic bottle) irrigation method

Years	Seasons	Humus %	N-NH4 mg/kg	N-NO2 mg/kg	N-NO3 mg/kg	P2O5 mg/kg	K2O mg/kg	PH
2000	Winter	1,19	75,5	25,3	152,0	34,4	165,2	6,5
	Spring	1,95	25,5	3,3	158,1	19,3	229,1	6,8
	Summer	1,62	14	1,2	130,9	20,7	511,6	7,2
	Autumn	1,97	22	6,1	152,1	22,6	398,3	8,1
Annual average		1,68	34,25	8,975	148,27	24,75	326,05	7,1
2001	Winter	2,587	65,4	23,9	149,7	39,7	215,7	6,8
	Spring	2,640	54,2	4,8	136,5	35,9	219,1	7,0
	Summer	2,085	41,9	2,7	127,8	24,6	312,9	7,2
	Autumn	2,256	34,5	6,9	145,3	21,5	299,5	7,4
Annual average		2,392	49,0	9,57	139,8	30,42	261,8	7,1
2002	Winter	2,881	30,4	14,1	152,1	41,7	345,4	7,6
	Spring	2,417	64,0	8,6	152,1	33,1	388,2	7,4
	Summer	2,783	28,0	6,6	124,0	14,2	324,9	7,0
	Autumn	2,571	29,1	4,7	120,3	10,1	278,0	7,1
Annual average		2,688	36,3	8,5	137,1	24,7	334,2	7,2

The weighted average content of chlorine ions for three years decreased 1.8 times, sulfate 1.5 times, calcium and magnesium increased, manganese steadily decreased to 62.2 mg/kg The weighted average content of copper and iron increased respectively 1.4-9.5 mg/kg and 1.3-4.5 mg/kg (Table 5.4.10.).

Thus, when using irrigation system made of plastic bottles (household waste), irrigation water is delivered inside the soil directly to the roots of plants and will save 4 times less water than furrow irrigation, good conditions for soil microorganisms are created. Labor resources are saved. Soil fertility increases steadily. Ammonium and nitrate forms of nitrogen are mainly accumulated in winter and spring seasons of the year, and during vegetation period their amount due to plant nutrition decreases.

Thus, when using irrigation system made of plastic bottles (household waste), irrigation water is delivered inside the soil directly to the roots of plants and will save 4 times less water than furrow irrigation, good conditions for soil microorganisms are created. Labor resources are saved. Soil fertility increases steadily. Ammonium and nitrate forms of nitrogen are mainly accumulated in winter and spring seasons of the year, and during vegetation period their amount due to plant nutrition decreases.

Table 5.4.10.

Seasonal and annual changes of chemical substances content (mg/kg) in 0-30 cm layer under in-soil irrigation method (plastic bottle)

Years	Seasons	CL mg/kg	SO4 mg/kg	Ca mg/kg	Mg mg/kg	Mn mg/kg	Cu mg/kg	Fe mg/kg
2000	Winter	2000	205,2	1920	336,0	66,8	2,4	2,1
	Spring	1400	344,2	1008	499,3	63,6	1,8	0,6
	Summer	1800	344,9	840	408,0	99,7	1,0	0,6
	Autumn	920	344,2	2256	347,4	99,7	0,5	2,0
Annual average		1530	309,6	1506	397,6	82,4	1,4	1,3
2001	Winter	1800	245,5	1830,0	324,0	76,2	2,7	1,9
	Spring	1650	367,2	995,0	465,0	60,5	1,9	0,5
	Summer	1900	370,3	865,0	409,0	93,4	1,2	0,65
	Autumn	2100	370,5	2143,0	365,0	95,0	0,9	1,75
Annual average		1862,5	338,3	1458,2	390,75	81,27	1,67	1,2
2002	Winter	1400	344,2	3200	584,1	99,7	6,2	2,84
	Spring	820	205,2	2459	447,3	56,4	7,4	2,21
	Summer	600	196,4	2435	414,8	44,7	6,2	2,2
	Autumn	640	192,1	1842	416,0	48,0	6,2	3,8
Annual average		865,0	183,69	2484,0	465,55	62,2	6,5	2,76

Thus, different irrigation methods differently affect the fertility of irrigated meadow alluvial saline soils, for example, in rain-fed irrigation method the content of gross humus in spring reaches 3.5% in other irrigation methods (furrow, drip, soaking hose, plastic bottles) there is a tendency of humus accumulation against the background of biointensive technologies (Fig. 11).

The content of nitrate nitrogen, mobile phosphorus and potassium increases and decreases several times by seasons of the year (Fig.12,13,14).

Chapter 6. Influence of different irrigation methods and biointensive technology on organic matter content of irrigated meadow alluvial saline soils

Preservation and increase of soil fertility, improvement of their ameliorative condition in the Lower Amudarya, where soil cover has been strongly degraded in recent years as a result of the Aral Sea desiccation, is very urgent.

In the current situation, irrigated soils of the region have significantly lost their fertility. To prevent these negative phenomena, along with reclamation measures, it is necessary to regulate the state of organic part of irrigated soils. As it is known, organic matter is one of the main criteria of soil fertility, many basic properties and soil regimes are closely related to it.

Fundamental works of major scientists on soil organic matter by I.V. Tyurin, M.M. Kononova, V.V. It was shown by I.V. Tyurin, M.M. Kononova, V.V. Ponomareva, L.N. Aleksandrova that the intensity of humus formation process is a function of bioclimatic and in-soil conditions (Orlov et al., 1987). Numerous studies of humus state of irrigated soils of the republic have revealed that humification processes are influenced by anthropogenic impacts, including irrigation (Ryzhov, Tashkuziev, 1976; Tashkuziev, 1996; Zakirov, Tashkuziev et al., 2000).

At present, science and practice of farming show that intensification of agricultural production can lead to deterioration of humus state of soils, especially in arid zone. This fact is most of all traced in connection with wide development of irrigation meliorations and under irrigation conditions under high irrigation water rates.

Humus mineralization in irrigated soils can reach large sizes, which can lead to prevalence of decomposition processes over accumulation processes and, in the absence of regulation, can greatly change the ecological situation in irrigated areas. In this regard, the study of transformation processes in the soil of composts from household waste, organic residues of various origins, aimed at enriching the soil with organic matter, is one of the most important tasks of practical soil science for the conditions of irrigated soils of arid zone.

We have studied the influence of biointensive and water-saving technologies on the background of organic fertilizers on the dynamics of changes in the content of organic matter and its qualitative composition in irrigated saline meadow alluvial soils, aimed at increasing their fertility and reducing salinization.

Irrigated meadow alluvial saline soils of the Lower Amudarya River served as objects of research. In establishing the influence of biointensive technology and different methods of irrigation on soil fertility and plant productivity, stationary methods were used by setting field plot experiments with composting and mulch application for vegetable crops under different methods of irrigation. Duration of experience - 2000-2002, after the action - 2003-2004. Repetition of experiments - 4^x multiple. The total area - 980 m^2. Furrow, rain, drip irrigation methods were used, as well as irrigation with the use of soaking hose and plastic bottles. The humus content in the soil was determined according to the method of I.V. Tyurin in modification of Simakov, fractional-group composition of humus - according to the scheme of I.V. Tyurin in modification of V.V. Ponomareva and T.A. Ponomareva. Ponomareva and T.A. Plotnikova (1966).

As it was mentioned above, field experiments for vegetable crops under different methods of irrigation with application of organic fertilizers annually at the rate of 15 t/ha in the form of compost, siderate and mulch were conducted in the period 2000-2002 and study of the effect of these fertilizers on changes in humus content and soil fertility were conducted in 2003-2004.

In this connection, in order to reveal the changes in the content and composition of soil organic matter during the period of research we determined the initial humus content in soil samples taken in the fall of 1999, i.e. before the experiments were laid and its content at the end of the experiment in the fall of 2004 (Table 6.1).

As can be seen from the data presented in Table 6.1, the applied technologies led to a significant change in the content of total humus in the soil for three years of the experiment and for the following two years of its after-effect. The greatest increase in humus content was observed in the upper 0-20cm, 20-50cm layers of soil, where humus accretion is about 2.074 and 1.783% or 3-3.1 times, respectively. Increase of humus content was also observed in lower soil horizons, also at the depth of 50 cm, but slightly weaker - 1.2 times. In the whole layer 0-120 cm the increase of humus content more than two times was noted. This shows a very positive impact of applied technologies on enrichment of soil with organic matter and, consequently, on increasing its fertility.

Table 6.1.

Changes in total humus content under water-saving and biointensive technology application

Source soil, R.1. 16.09.1999.				At the end of experiment R.2. 15.09.2004.				augmentation	
Gubin, cf.	Humus, %	In layer, cm	Humus, %	Depth, cm.	Humus, %	In layer, cm	Humus, %	%	Difference
0-10	1,043	0-20	1,034	0-10	3,752	0-20	3,108	2,074	3,0
10-20	1,025	0-50	0,859	10-45	2,464	0-50	2,642	1,783	3,1
20-47	0,745	50-120	0,697	45-60	1,670	50-120	0,844	0,147	1,2
47-80	0,715	0-120	0,765	60-70	0,879	0-120	1,593	0,828	2,1
80-120	0,683			70-120	0,672				

Establishing the influence of applied technologies is important for the composition of soil organic matter. In this regard, we studied the fractional-group composition of humus of soils of the initial state and for the elapsed 5-year period of the experiment. The results of these studies are summarized in Tables 6.2. - 6.4. These data show that the applied technology leads to a noticeable change in the soil content of humic acids and especially insoluble residue.

To clarify this, let us first of all dwell on the changes that occurred in the composition of individual fractions of soil humic acids (Table 6.2).

From the data of Table 6.2 it can be seen that in comparison with the initial soil, in genetic horizons of soil after 5 years of application of the new technology, a marked increase in the absolute amount of carbon (in % to the weight of soil) in all fractions, both humic acids and fulvic acids. This is due to a significant increase in the content of total humus throughout the soil profile as a result of the application of biointensive technology aimed at enriching the soil with organic matter.

Table 6.2.

Fractional composition of humus

Soil	Depth, cm	Humic acids, fraction			Fulvic Acids, Fraction			
		1	2	3	1a	1	2	3
Nukus No. 20 m/district 16.09. 1999 P - 1	0-10	13,98 / 0,0846	6,35 / 0,0384	14,30 / 0,0865	4,48 / 0,0271	3,90 / 0,0236	16,43 / 0,0994	13,36 / 0,0808
	10-20	11,63 / 0,0692	6,97 / 0,0415	12,47 / 0,0742	4,13 / 0,0246	4,91 / 0,0292	15,36 / 0,0914	12,34 / 0,0734
	20-47	12,45 / 0,0538	10,32 / 0,0446	14,31 / 0,0618	4,56 / 0,0197	7,20 / 0,0311	15,58 / 0,0673	15,30 / 0,0661
	47-80	8,87 / 0,0368	11,88 / 0,0493	11,90 / 0,0494	3,57 / 0,0148	9,81 / 0,0407	13,33 / 0,0553	16,55 / 0,0687
	80-120	3,89 / 0,0154	8,54 / 0,0338	9,37 / 0,0371	311 / 0,0123	8,54 / 0,0338	11,42 / 0,0452	10,50 / 0,0416
Nukus No. 20 m/r-n 15.09. 2004 R - 2	0-10	6,01 / 0,1307	4,73 / 0,1030	8,52 / 0,1854	1,70 / 0,0369	2,19 / 0,0477	7,42 / 0,1614	5,05 / 0,1098
	10-45	6,46 / 0,0923	5,59 / 0,0799	10,38 / 0,1483	1,90 / 0,0271	4,56 / 0,0651	10,25 / 0,1465	8,21 / 0,1174
	45-60	6,35 / 0,0615	8,88 / 0,0861	12,76 / 0,1236	2,28 / 0,0221	8,51 / 0,0825	12,82 / 0,1242	12,63 / 0,1224
	60-70	10,55 / 0,0538	15,98 / 0,0815	16,96 / 0,0865	4,33 / 0,0221	12,86 / 0,0656	14,63 / 0,0746	17,76 / 0,0906
	70-120	5,92 / 0,0231	13,31 / 0,0753	15,85 / 0,0618	4,41 / 0,0172	18,05 / 0,0704	12,23 / 0,477	16,95 / 0,0661

Note: in numerator, the amount of carbon of fractions in % to total carbon, in denominator, in % to soil weight

However, there is a noticeable decrease in the relative amount (in % to total carbon) of humic acids of fractions 1 and 2 and fulvic acids of fractions 1-a, 2 and 3 in the first half-meter horizon of soils after the experiment. Downwards these indices remain almost equal or with some increase in the relative amount of humic acids of these fractions in soils of section 2, where biointensive technology was applied. The decrease in the relative amount (in % to total carbon) of humus acids of these fractions in the upper horizons is due to excessive increase of organic carbon content, which does not allow its full fixation by calcium and magnesium cations and semithin oxides in these layers. At the same time, a significant part of newly formed organic carbon is apparently bound by the colloidal part of the soil.

It is of interest how the group composition of humus in separate genetic horizons of soil develops as a result of three-year application of the new technology (Table 6.3).

As can be seen from the presented data in Table 6.3, in accordance with the increase in horizons of organic carbon content (carbon in soil), there is a marked increase in the absolute content of both humic acids and fulvic acids in them. For example, in horizons 0-10, 10-20 and 20-27 cm of the original soil, the absolute amount of humic and fulvic acids ranges from 0.1602-0.2095 % and 0.1842-0.2309 % of the soil weight, respectively.

In similar soil horizons after application of technologies (also in soils of experimental plot) these indices make 0.3205-0.4191 % and 0.35610.3558 % accordingly, that exceeds in 1.5-2 times in comparison with initial soil. A significant increase in the absolute amount of non-hydrolyzable residue was also noted.

Humic in the upper 0-60 cm horizons of the soils of the experimental field - 0.3466-1.4015 %, against 0.1000-0.1915 %. Such an increase in the absolute amount of carbon of hydrolyzed humus substances in the soils of the experimental plot is associated with a significant increase in the amount of total organic carbon in the upper horizons. In the underlying horizons of this soil a small increase in soil carbon is traced in comparison with similar horizons of the initial soil. In accordance with this, in the lower layers of the soil of the experimental plot there is an increase in the absolute amount of humic and fulvic acids in 1.3-2 times in comparison with similar horizons of the original soil.

It should be noted that the increase in the absolute amount of carbon of humus acids is not always accompanied by an increase in the relative amount of humic acids and fulvic acids so, as noted above, with a significant increase in the content of organic carbon in the soil, its binding by calcium cations and semifluoric oxides occurs partially and the remaining part is apparently bound to the colloidal part of the soil, is also part of fine particles and is bound by clay minerals. In this regard, in the upper 0-60 cm layers of the soil of the experimental plot the content of non-hydrolyzable residue exceeds 5-6 times in comparison with the original soil.

Table 6.3.

Group composition of humus

Soil, option	Depth, cm	C in soil, %	Humic acids, Cgc	Fulvo acids, Sfc	Non-hydrolyzable residue	Attitude	
						Sgc Sfc	Sg Sng
Nukus No. 20 m/r-n	0-10	0,605	34,63 0,2095	38,16 0,2309	27,21 0,1646	0,91	2,68
	10-20	0,595	31,07 0,1849	36,74 0,2186	32,19 0,1915	0,85	2,11

16.09. 1999 R - 1	20-47	0,432	37,08 0,1602	42,64 0,1842	20,28 0,0876	0,87	3,93
	47-80	0,415	32,65 0,1355	43,26 0,1795	24,09 0,1000	0,75	3,15
	80-120	0,396	21,80 0,863	33,56 0,1329	44,65 0,1768	0,65	1,24
Nukus No. 20 m/r-n 15.09. 2004 P - 2	0-10	2,176	19,26 0,4191	16,35 0,3558	64,39 1,4011	1,18	0,55
	10,45	1,429	22,43 0,3205	24,92 0,3561	52,65 0,7524	0,90	0,90
	45-60	0,969	27,99 0,2712	36,24 0,3512	35,77 0,3466	0,77	1,80
	60-70	0,510	43,49 0,2218	49,59 0,2529	6,92 0,0353	0,88	13,45
	70-120	0,390	41,08 0,1602	51,64 0,2014	7,28 0,0284	0,80	12,73

Note: in numerator - amount of carbon of separate groups of humus substances in % to total carbon, in denominator - in % to soil weight.
in the numerator is the amount of carbon of separate groups of humus substances in % to total carbon, in the denominator - in % to soil weight.

From the presented data of Table 6.3 it is obvious that except for 0-10 cm layer of the experimental plot soil, in all horizons of the studied soils the ratio Cgc : Cfc is less than one (0,75-0,91) m humus type refers to humate-fulvate type, also remains almost unchanged with some approximation to fulvate-humate type in the soils of the experimental plot. The applied technology contributed to the change of humus type in the horizon 0-10 cm, where the ratio Cgc : Cfc = 1.18 m by humus type belongs to fulvate-humate type. It is in this layer that the transition of quantitative change to qualitative change was noted.

In general, we can conclude that the applied biointensive technology along with the increase in the genetic horizons of the studied soils of organic carbon in comparison with the original soil, contributed to a noticeable change in the composition of hydrolyzable humus substances and non-hydrolyzable part of humus. Generalization of similar materials obtained by us on humus composition was reflected in Table 6.4.

The data from this table clearly shows a significant increase in the content of total soil organic carbon in the root-inhabited layer (0-70 cm) of the soil of the experimental plot, where the biointensive technology was applied with the use of organic fertilizers at a rate of 15 t/ha for three years in the form of compost, siderate and mulch. Especially the increase in the amount of soil organic carbon is very strong (3 m more times) in the first half-meter layer. In accordance with this and traces noticeable changes in the fractional and group composition of humus in the genetic horizons of the root-inhabited layer of soil (0-60-70cm), which was noted above when discussing the data in Tables 6.2 and 6.3.

At application of bio-intensive technology after years of experiment also reveals some decrease in the relative amount of carbon of all fractions of humic acids of the upper 0-60 cm horizons and a noticeable increase in the underlying soil layers. Such changes are traced in relation to fractions of fulvic acids, but somewhat weaker. Consequently, the applied technology has noticeably changed the composition of humic acids humus, the most important part in relation to the indicator of humus state of the soil, since the intensity of organic matter humification is determined precisely by the content of humic acid fractions in the soil.

Table 6.4.

Fractional-group composition of humus in soil, % to total carbon.

Soil, option	Depth, cm	C in soil, %	Humic acids,				Fulvic acids					Sgc Sfc	Hydro-lysed substances, C_d	Non-hydrolyzed residue, Cng
			1	2	3	Amount	1a	1	2	3	Amount			
	0-10	0,605	13,48	6,35	14,30	34,63	4,48	3,90	16,43	13,36	38,16	0,91	72,79	27,21
	10-20	0,595	11,63	6,97	12,47	31,07	4,13	4,91	15,36	12,34	3674	0,85	67,81	32,19
Nukus. No. 20 m/district 16.09. 1999 P- 1	20-47	0,432	12,45	10,32	14,31	37,08	4,56	7,20	15,58	15,30	42,64	0,87	79,72	20,28
	47-80	0,415	8,87	11,88	11,90	32,65	3,57	9,81	13,33	16,55	43,26	0,75	75,91	24,09
	80-120	0,396	3,89	8,54	9,37	21,80	3,11	8,54	11,42	10,50	33,56	0,65	55,35	44,65
	0-10	2,176	6,01	4,73	8,52	19,26	1,70	2,19	7,42	5,05	16,35	1,18	35,61	64,39
	10,45	1,429	6,46	5,59	10,38	22,43	1,90	4,56	10,25	8,21	24,92	0,90	47,35	52,65
Nukus. No. 20 m/r-n 15.09. 2004 R-2	45-60	0,969	6,35	8,88	12,76	27,99	2,28	8,51	12,82	12,63	36,24	0,77	64,23	35,77
	60-70	0,510	10,55	15,98	16,96	43,49	4,33	12,86	14,63	17,76	49,59	0,88	93,08	6,92
	70-120	0,390	5,92	19,31	15,85	41,08	4,41	18,05	12,23	16,95	51,64	0,80	92,72	7,28

It can also be noted that the applied technology caused noticeable changes in this root-inhabited layer of soil and in relation to the hydrolyzable part of humus acids. As we can see in genetic horizons in the 0-60 cm layer of soil of the experimental plot, the amount of hydrolyzable humus substances decreases by 1.5-2 times and increases strongly in the underlying soil horizons.

From the above, the following conclusions can be drawn:

1. The applied water-saving and biointensive technologies in crop cultivation in irrigated meadow alluvial soils of the Amudarya lower reaches have a positive impact on the accumulation of organic matter in the soil, especially in the root-inhabited layer.
2. The increase in soil organic matter content is stable, which is maintained in the next

two years of the consequences of the measures taken to improve soil fertility under the new technology.

3. The applied technologies aimed at improving soil fertility, obtaining high ecologically clean bioproducts have a very positive impact on the qualitative composition of humus and in general on the humus state of the soil.

Chapter 7. Effect of different irrigation methods and biointensive technology on tomato and eggplant yields

In 2000, from different irrigation methods and norms, an increase in tomato yield from 39.2 to 46.2c/ha was obtained, while in the control the yield was 290.5c/ha (Table 7.1).

Yield data showed that reducing the rate of irrigation up to 3000 and 1500m^3 in rain and drip irrigation methods gave negative results, and in the case of in-soil irrigation method, despite the 4 times reduction of irrigation rates gave positive results as tomato yield increased from 39.2 to 46.2 c/ha.

Reduction of irrigation rate from 6000 to 1500 $m^{(3)/ha}$ under rainfall and drip irrigation methods resulted in reduction of tomato yield from 71.6 to 88.6 kg/ha (Table 7.1).

The results of the 2000 experience show that the biointensive method of soil tillage and the use of in-soil irrigation provides high yields despite the reduction of irrigation rates by 4 times.

From the application of drip irrigation, which reduced irrigation rates 4 times, but no increase in tomato yield was obtained.

In 2001, from different methods of irrigation rates, an increase in tomato yield from 24.1 to 28.6 c/ha was obtained, while in the control the yield was 310.7 c/ha.

Yield data showed that reduction of irrigation rate at rain and drip methods gave negative results i.e. yield from 95.4 to 87.6 c/ha decreased.

At in-soil method of irrigation tomato yield increased from 24.1 to 28.6 c/ha and despite the reduction of irrigation rate by 4 times gave positive results.

The results of the 2001 experiment show that the biointensive method of soil tillage and the use of in-soil irrigation provided a higher yield than furrow irrigation, which used 6000 $m^{(3)\backslash ha\ of}$ water.

The results of the analysis of yield data of the experiment for 2001 show that under rain and drip irrigation does not indicate a significant effect on increasing the yield of tomato, except within the soil method of irrigation, which contributed to a noticeable increase in yield.

Application of plastic bottles only as in-soil irrigation contributed to saving of irrigation water 4 times, which exceeded the control by 109.2%. Application of rainwater irrigation method did not result in yield increase, despite saving irrigation water 2 times.

In 2002 from different methods of irrigation rates was obtained an increase in tomato yield from 31.5 to 41.4 c/ha. And in the control variant the yield was 313.5 c/ha.

Yield data showed that as in previous years, the reduction of irrigation rates in rain and drip irrigation methods gave negative results as tomato yields decreased from 103.6 to 98.6 c/ha. In the in-soil irrigation method the yield increased by 110%.

Table.7.1.

Tomato yields depending on different methods and rates of irrigation (c/ha.)

Options	Repetition				Medium	Yield increase
	I	II	III	IV		
			2000 NDS05- 0.91 c/ha			
1	289,2	290,3	287,9	294,6	290,5	0
2	198,5	200,0	203,0	205,5	201,9	- 88,6
3	214,9	213,1	223,4	221,4	218,9	- 71,6
4	327,9	326,8	331,5	332,6	329,7	+ 39,2
5	334,7	332,9	338,2	341,0	336,7	+ 46,2
			2001 NDS05- 0.52 c/ha			
1	311,0	309,2	310,0	312,6	310,7	0
2	313,7	316,0	314,0	317,5	215,3	- 95,4
3	218,2	223,0	222,9	225,0	223,1	- 87,6
4	329,7	329,9	331,1	334,2	334,8	+ 24,1
5	337,9	338,0	339,9	341,4	339,3	+ 28,6
			2002 NDS05- 0.24 c/ha			
1	312,9	313,6	314,7	312,8	313,5	0
2	209,0	210,1	209,3	210,4	209,7	- 103,6
3	213,9	215,1	214,9	215,7	214,9	- 98,6
4	343,9	346,3	343,9	345,1	344,8	+ 31,3
5	355,0	354,8	353,7	356,1	354,9	+ 41,4

The results of the 2002 experiment show that the application of biointensive tillage methods and water-saving technologies provided a higher yield than furrow irrigation, which uses 6000 m(3)/ha of water during the growing season.

Our study for 2002 shows that under conditions of rain and drip irrigation does not have a significant effect on increasing tomato yields, except for in-soil irrigation, which contributed to increased yields. Application as in-soil irrigation soaking hoses and plastic bottles contributed to saving irrigation water 4 times, exceeded the control (6000 m^3) by 110 %.

In 2000, different irrigation methods and rates gave an increase in eggplant yield from 0.9 to 12.6 kg/ha, while in the control the yield was 763.3 kg/ha (Table 7.2).

Yield data showed that reducing the rate of irrigation to 3000 and 1500m^3 in the rain and drip irrigation methods gave negative results, and with in-soil irrigation method despite the 4 times reduction of irrigation rates gave positive results because the yield of eggplant increased from 9.5 to 12.6 c/ha.

Reducing the irrigation rate from 6000 to 1500 m(3)/ha under rain-fed and drip irrigation methods led to a decrease in eggplant yield from 0.7 to 12.2 kg/ha (Table 7.2).

The results of the 2000 experience show that the biointensive method of soil tillage and the use of in-soil irrigation provides high yields despite the reduction of irrigation rates by 4 times.

From the application of drip irrigation, which reduced irrigation rates 4 times, but no increase in eggplant yield was obtained.

In 2001, from different methods of irrigation rates an increase in eggplant yield from 5.2 to 12.6 c/ha was obtained, while in the control the yield was 775.3 c/ha.

Yield data showed that reduction of irrigation rate at rain and drip methods gave negative results i.e. yield decreased from 3.8 to 1.4 c/ha.

At the in-soil method of irrigation yield of eggplant increased from 5.2 to 12.6 c/ha and despite the reduction of irrigation rate by 4 times gave positive results.

The results of the 2001 experiment show that the biointensive method of soil tillage and the use of in-soil irrigation provided a higher yield than furrow irrigation, which used 6000 m(3)\ha of water.

The results of the analysis of yield data of the experiment for 2001 show that under the conditions of rain and drip irrigation does not indicate a significant effect on increasing the yield of eggplant, except within the soil method of irrigation, which contributed to a noticeable increase in yield.

Application only as in-soil irrigation plastic bottles contributed to the saving of irrigation water 4 times, which exceeded the control by 102.0% From the use of rainwater irrigation method was not obtained yield increase, despite the saving of irrigation water 2 times.

In 2002 from different methods of irrigation rates was obtained an increase in eggplant yield from 4.7 to 9.5 c/ha. And in the control variant the yield was 790.0 c/ha.

Yield data showed that as in previous years, reducing the rate of irrigation with rain and drip irrigation methods gave negative results because the yield of eggplant decreased from 12.2 to 9.9 c/ha. In the in-soil method of irrigation yield increased by 102%.

Table 7.2.

Eggplant yields depending on different irrigation methods and rates (c/ha)

Variant	Repetition				Medium	Yield increase
	I	II	III	IV		
			2000	HCP_{05}-0.48 c/ha		
1	763,3	760,9	759,0	761,0	761,0	0
2	758,9	759,7	760,5	762,1	760,3	- 0,7
3	757,9	762,3	758,9	760,1	759,8	- 1,2
4	760,3	759,9	764,3	763,1	761,9	+ 0,9
5	761,9	762,6	763,2	766,1	763,5	+ 2,5
			2001	NDS_{05} - 0.36 c/ha		
1	774,6	775,0	774,9	776,7	775,3	0
2	773,3	772,3	775,1	774,9	773,9	- 1,4
3	769,9	773,0	770,8	772,3	771,5	- 3,8
4	779,4	780,2	781,1	781,3	780,5	+ 5,2
5	786,9	785,8	790,6	788,3	787,9	+ 12,6

			2002	NDS05 - 0.27 c/ha		
1	789,5	790,9	791,9	791,3	790,0	0
2	781,9	778,5	779,0	781,0	780,1	- 9,9
3	777,1	776,9	778,7	778,5	777,8	- 12,2
4	793,6	795,1	794,9	795,2	794,7	+ 4,7
5	798,8	799,4	799,4	800,1	799,5	+ 9,5

The results of the 2002 experiment show that the application of biointensive methods of soil tillage and the use of water-saving technologies provided a high yield of eggplant than furrow irrigation, which in the growing season used 6000 m$^{(3)\backslash \text{ha of}}$ water.

Our study for 2002 shows that under conditions of rain and drip irrigation does not have a significant effect on increasing eggplant yields, except for in-soil irrigation, which contributed to increased yields. Application as in-soil irrigation impregnating hoses and plastic bottles contributed to saving irrigation water 4 times, exceeded the control (6000 m^3) by 102 %.

Conclusions

The results of studies on the influence of different irrigation methods and biointensive methods on fertility and degree of salinization of irrigated soils allow us to draw the following conclusions:

1. In low-water years, the use of drinking and irrigation water for soil irrigation had a significant impact on the increase of salinization of irrigated soils, as a result of which the source of salt accumulation in the soil was not only soil-soil water, but also irrigation water and tap water supplied for irrigation of vegetable crops.

2. It is established that when applied to vegetable crops, furrow irrigation in the conditions of annual application of organic fertilizers in the rate of 15t/ha in the form of compost, siderate and mulch there is an increase in humus in comparison with a similar indicator in 1.5 times.

3. When growing vegetable crops under biointensive technologies using rain-fed irrigation with irrigation rate of $3000m^{(3)\backslash ha}$, there is an increase in humus content compared to the initial state of soils in 2.0-2.5 times.

4. When using drip irrigation method under vegetable crops on organic background in arable layer of soils humus content increases 1.7 times, ammonium nitrogen-2 times. Average weighted content of nitrate nitrogen, mobile forms of phosphorus and potassium decreases.

5. When using soaker hoses for in-soil irrigation, the content of humus, nitrate nitrogen, potassium, calcium, magnesium and copper increases. Nitrate nitrogen, mobile phosphorus and iron remain at the same level.

6. When using the irrigation system prepared from plastic bottles, irrigation water is delivered into the soil directly to the roots and allows saving irrigation water 4 times ($1500m^3$/ha) than furrow irrigation, good conditions for soil microflora are created. Soil fertility increases steadily. Ammonium and nitrate forms of nitrogen are mainly accumulated in the winter-spring season of the year, and during the growing season their amount decreases due to assimilation by plants. Of water-soluble salts chlorine- 1.8 times, sulfate- 1.5 times decreases.

7. When using in-soil irrigation method (soaking hoses, plastic bottles) during the growing season on irrigated meadow alluvial saline soils on the background of biointensive technology in the arable layer of soils there is a decrease in salt content compared to furrow irrigation.

8. The applied water-saving and bio-intensive technologies for crop cultivation in irrigated meadow alluvial saline soils of the Amudarya lower reaches have a positive impact on fertility of saline soils of the Amudarya lower reaches, contributes to the accumulation of organic matter in the soil, especially in the root-inhabited layer.

9. The increase in soil organic matter content is stable, which is maintained in the following two years after the measures taken to improve soil fertility using the new technology.

10. The applied technologies aimed at increasing soil fertility, obtaining high,

ecologically clean bioproduct has a positive impact on the qualitative composition of humus, and in general on the humus state of irrigated meadow alluvial saline soils.

11. Application of bio-intensive methods of soil tillage provided high yields than furrow irrigation. In-soil irrigation methods promoted saving of irrigated water 4 times, and tomato yield increased by 110%, and eggplant yield increased by 102%.

Proposals to Production

The use of bio-intensive technologies on the background of organic fertilizers, as well as the introduction of in-soil irrigation using plastic bottles on irrigated meadow alluvial saline soils contribute to improving soil fertility and reducing soil salinization.

In conditions of ecological crisis on irrigated meadow alluvial saline soils for obtaining high and quality yields of vegetable crops it is recommended to apply biointensive technologies on the background of organic matter (15t/ha) with the above mentioned method of in-soil irrigation.

Literature

1. Abdullaev S.A., Bairov A.J. Current state of irrigated soils of Uzbekistan and priorities in scientific support of conservation and improvement of their fertility // Bulletin of Agrarian Science of Uzbekistan, № 1(7), 2002, p.66-69.
2. Alekseev E.K. Green fertilizer in the Non-Black Earth strip. M. Selkhozizdat. 1959. 320 c.
3. Bairov A.J. Problems of creation of automated information systems in soil science. // Bulletin of Agrarian Science of Uzbekistan, No. 1, 2000, p.47-51.
4. Baimatova L.S., Slesareva L.N. Influence of crop rotation scheme, cotton monoculture and depth of basic tillage on water-physical properties of irrigated typical sierozems // Abstracts of the 1st delegate of the Congress of Soil Scientists of Uzbekistan. Congress of Soil Scientists of Uzbekistan. Tashkent.1990. C.39.
5. Balashev N.N., Zeman G.O. Vegetable growing Tashkent, "Ukutuvchi", 1981, 368 pp.
6. Balyabo N.K. Development and increase of fertility of solonetz soils. M. Selkhozizdat. 1961. C.120.
7. Batkaev J.Ya. Changing conditions of fertility of dark sierozems under the influence of agricultural crops and fertilizers. Dissertation for the degree of Candidate of Agricultural Sciences. Tashkent.1962. 17 c.
8. Bezdnina S.Ya. Influence of irrigation water quality on soil fertility // Theses of reports of the 111th Congress of Dokuchaevsky Society of Soil Scientists (July 11-15, 2000, Suzdal) Book 2. Moscow, 2000, p.211.
9. Belik V.F., Bondorenko G.L. Methodology of field experience in vegetable growing and melon growing. Moscow, Izd vo Kolos, 1979.
10. Belousov M. A. Influence of long-term application of organic and mineral fertilizers on the productive capacity of irrigated sierozems // Influence of long-term application of fertilizers on soil fertility and productivity of crop rotations.- Vyp. 1.-M.MSH OF THE USSR. 1960. C.366-383.
11. Borovsky V.M., Ablakov E.B., Kozhevnikov K.Y., Muravlinsky K.D. . Ancient Syr-Darya Delta and Northern Kyzylkums.vol.1, ANKazSSR Publishing House, Alma-Ata, 1959,420 p.
12. Borovskiy V.M., Pogrebinsky M.A. Ancient Syr-Darya Delta and Northern Kyzylkums, Vol.1, Publishing House of ANKazSSR, Alma-Ata, 1958,
13. Buriev H.Ch., Abdullaev A.G. Homestead vegetable garden. Tashkent, <Mekhnat>, 1986, 216 p.
14. Vasiliev V.A., Filippova N.V. Reference book on organic fertilizers.- 2nd edition, revision and supplement. Moscow. Rosagropromizdat, 1988. 255 c.
15. Vakhobov A., Mirzaev O. Zhuraev 3. Influence of biohumus on growth and development of cotton plant \\\ Bulletin of Agrarian Science of Uzbekistan, № 2(4), 2001, p.14-15.
16. Davranov K.D., Karshieva D.H. Role of microorganisms in soil health improvement, increase of their fertility and yield of agricultural plants \\\\ Bulletin of Agrarian Science

of Uzbekistan, № 4(10), 2002, p.64-67.
17 John Jevons *How to Grow More Vegetables....* Company< Pacific BVL Corporations 1993, 178 C
18.John Jevons "How to Grow More Vegetables". Second edition 1997, 224 pp.
19.John Jevons, Carol Cox "The organic vegetable garden" A guide to growing healthy soil and high yields on your plot. Ecology Action Group Illustrations by Sue Ellen Parkinson, 1999, 116 pp.
20.Dovban K.I. Wider introduction of sideration in intensive farming // Zemledeliye. 1990. № 12. C.32-34.
21.Yokubov M. Eggplant // Agriculture of Uzbekistan. 2003, No. 7, p.23.
22 Ermolova E. New tomato varieties // Agriculture of Uzbekistan. 2003, No. 7, pp.20-21.
23.Zhollibekov B. Changes in soil cover of Amudarya coastal delta under aridization. Nukus, "Bilim" Publishing House, 1991, 132.
24.Zhollibekov B. Changes in soil cover and landscapes of the Southern Priaralie in connection with anthropogenic impact. Nukus, Bilim Publishing House, 1995, 244 pp.
25.Zhollibekov B. Change of soil cover of irrigated part of Amudarya delta. Soil Science, No.7. 1994, pp.91-97.
26. Zhollibekov B. Monitoring of drinking water quality in the epicenter of the Aral catastrophe // Novosti Zdravookhranenie Informatsionny Bulletin, 5th issue, supplement to the journal "Zdravookhranie Kazakhstana", February 1997, pp. 8-9.
27.Zhollibekov B. Ways to achieve sustainable growth of agriculture of the republic // "Karakalpakstan"(public*)*. "Karakalpakstan" (socially - political Educational Journal), Nukus, 2000, No. 5-6, pp.
28.Zholybekov B., Zholybekov B.B. Nambar kalai tayarlanady? // Journal "Zhipek zholy-Silk Way," # 39-40, 20-September 2003, 4-5 pp.
29.Zhollibekov B., Ismaylov U.E.,Tleuov A., Turmanova O.K., Ismaylova R., Zhollibekov B. B. // Use of agricultural crops as biomeliorants in conditions of ecological crisis. // Bulletin of Agrarian Science of Uzbekistan, No. 4 (14), 2003, pp.63-68.
30.Zakirov T.S., Tashkuziev M.M., Ibragimov N.M., Niyazaliev B.I. On the results of research on humus formation in sulfur soils of Uzbekistan based on the study of decomposition of plant residues and application of fertilizers. \\ Bulletin of Agrarian Science of Uzbekistan, No. 1, 2000, pp. 52-55.
31.Zakirov T.S., Tashkuziev M.M., Ibragimov N.M., Niyazaliev B.I. On the results of studies of humus formation in gray soils of Uzbekistan based on the study of decomposition of plant residues and application of fertilizers. J. Uzbekistan agrar fani khabarnomasi, No.1, 2000, p. 52-55.
32.Zimina N.M. To the question about the influence of crop rotation, fertilizers and tillage on agrophysical properties of typical sierozem // Proceedings of SoyuzNIIKhI. Tashkent.

SAGU, 1955. C.44.

33.Zimina N.M. Change of physical properties of sierozems under the influence of cotton and alfalfa crop rotation and fertilizers // Soil Science. 1961. №8. C.470-480.

34.Ziyamukhamedov I.A., Ryzhov S.N. Organic matter of virgin and irrigated soils. Tashkent. Fan.1975.

35.Zueva V.I. (under general editorship) Vegetable growing laboratory-practical lessons. Tashkent, "Ukituvchi", 1980. 211c.

36.Imamova H.H., Sharipova H.T., Poberezskaya S.K. et al. Directionality of microbiological processes during composting of manure and chicken manure with Kyzylkum phosphorites // Bulletin of Agrarian Science of Uzbekistan, № 3(9),2002, p.30-33.

37.Irrigation of Uzbektistan, Vol. 111, "FAN" Uzbek SSR, Tashkent, 1979, 360 pp.

38.Ismailov U.E. Optimization of water regime under cotton irrigation with drainage water on meadow alluvial soils of Karakalpakstan// Abstract of thesis for the degree of Candidate of Agricultural Sciences. Tashkent, 1983, 20 p.

39.Ismailov U., Zhollibekov B. Crop rotation and fertility improvement // Agriculture of Uzbekistan. No.12, 2003, p.13.

40.Karimov O.T. Efficiency of biohumus application under cotton in conditions of typical gray soils. Abstract for the degree of Candidate of Agricultural Sciences, Tashkent, 2002, 23 pp.

41.Karimov O.T. Effectiveness of different doses of application of biohumus. Improvement of cultivation technology of agricultural crops in the Upper Volga region. Collection of articles, vol.3, Y Ivanovo, 2001, 145-146 p.

42.Karimov O.T. Effectiveness of "Biohumus" application rates for cotton in conditions of old-irrigated typical gray soils. gray soils. Bulletin VIUA,2001,№114, p.101.

43.Kachinsky N.A. Mechanical and microaggregate composition of soil, methods of its study. Izd. of the Academy of Sciences of the USSR, 1958.

44.Kachinsky N.A. Soil, its properties and life. M. ACADEMY OF SCIENCES OF THE USSR. 1956. C. 127.

45.Kachinsky N.A. Soil Structure. MOSCOW STATE UNIVERSITY. 1963. C. 100.

46.Kovda V.A. Origin and regime of saline soils. Vol. 1(1945),vol. 11(1947), Izd. of the Academy of Sciences of the USSR, M.-L

47.Kozlovsky F.I. Evolution of arable soils as a subject of genetic-geographical soil science. In prn. Teoriya i metody izucheniya zavisimosti sobstvennogo koverova (Theory and methods of studying soil cover), Moscow, GEOS. 2003. pp. 451-463.

48.Kozlovsky F.I., Pankova E.I. On the methodology of studying the salt regime of soils (by the example of some types of soils of the Jizzak steppe) // in the collection, Irrigated soils and methods of their study, Tashkent, 1976.

49.Kononova M.M. Humus of the main types of soils of the USSR, its nature and ways of formation // Soil Science. 1960. № 3 C. 1-11.

50.Kononova M.M. Organic matter of soil. M. ACADEMY OF SCIENCES OF THE USSR. 1963. 314 c.

51.Konstantinova L.G., Kurbanov A.B., Atanazarov K.M., Absattarov N.A. Quality of drinking water, health status of the population and forecast of morbidity of the population of the Republic of Karakalpakstan. Materials International Seminar "Ecological Factors and Maternal/Child Health in the Aral Sea Crisis Region" Publishing House "FAN", Academy of Sciences of Uzbekistan. 2001, C. 8795.

52.Korsunova D-C. Microbiological transformation of matter Fertilizers and plant residues in chestnut soils of Buryatia// Theses of reports of the 111th Congress of the Dokuchaev Society of Soil Scientists (July 11-15, 2000, Suzdal) Book 2. Москва,2000,стр.133-134.

53.Kudratov T. Ekubov Sh. Dare suvlari sifati// Uzbekistan Kishlok Khojaligi, 2003, №6, 24 pp.

54.Kuziev R.K. Problems of soil fertility in Uzbekistan. Tuprokshunoslar va agrochemiegarlar 111-chi kuroltoyida maruzalari va tezislari. Toshkent.2000.

55.Kuziev R.K. Modern state of irrigated soils and problems of its protection and reproduction. \\ Bulletin of Agrarian Science of Uzbekistan, No.1, 2000, p.43-46.

5b.Kuziev R.K., Tashkuziev M.M., Ziyamukhamedov I.A. Concept of biogeochemical technology basis for optimization of chemical state and real increase of soil fertility. \\ Bulletin of Agrarian Science of Uzbekistan, № 2(8), 2002, p.55-58.

57.Kuziev R.K., Toshkuziev M.M., Ziyamukhamedov I.A. Tuprok khosil bulish zhariyoni va unimdorligini oshirishning biogeokimyoviy asosi. Er resurslariridan akilona poydolanis va tuproklarni mukhofazalash. Ilmiy Tuplam. Toshkent.2001.

58.Cook J.W. Regulation of soil fertility. M. Kolos, 1970 520 p.

59.Kurvantaev R. Optimization and regulation of agrophysical condition of irrigated soils of desert zone of Uzbekistan. Tashkent, 2000, p. 45.

b0.Kurvantaev R., Kholberdiyeva M.R. Influence of mulching on soil water regime at sowing on ridges\\\ Vestnik of Agrarian Science of Uzbekistan, № 2(4), 2001, p.44-45.

b1.Loshakov V.G., Ivanova S.F. Fertility of sod-podzolic soil and productivity of crop rotations with different saturation with grain crops // Izv.TSKhA. 1975. № 3 C.41-51.

б2. Methods of agrochemical, agrophysical and microbiological studies in irrigated cotton areas (3rd revised and supplemented edition). Tashkent 1963.

b3.Methodology of field and vegetative experiments with cotton // Tashkent, SOYUZNIKHI, 1973.

b4.Minashina N.G. Irrigated desert soils and their amelioration. Moscow. Izd vo "Kolos".1974.3b5 p.

b5.Mirzaev M.M., Mirzaev M. Sideration is an effective method of increasing soil fertility in the gardens of Uzbekistan. // Bulletin of Agrarian Science of Uzbekistan, № 2,2000, p.59-b1.

bb.Mironenkov A. Water supply problems in Priaralie and action program. Materials of the international seminar "Ecological factors and maternal/child health in the Aral Sea region". Aral crisis" Izd. FAN Academy of Sciences of the Republic of Uzbekistan. 2001, P. b-11.

b7.Mukhamedjanov M.V. Take care of the land, multiply its fertility. Zemledeliye. 198б. № 1. C. 3-б.

b8.Nurzhanov K.R. Qualitative change of salt composition in soil depending on irrigation regime of cotton and corn. -Proceedings of SoyuzNIIKhI, issue 53, 1983.

b9.Nurzhanov K.R. Comparative water consumption and irrigation regime of cotton and corn on meadow soils of Karakalpakia. Tashkent, 1984, 19 p.

70.Ovezova K. Water-saving technology of cultivation of melon crops. Abstract of dissertation for the degree of Doctor of Agricultural Sciences. Tashkent, 1991, 44 p.

71.Orlov D.S., Baranovskaya V.A., Okolelova A.A. Organic matter of steppe soils of the Volga region and its transformation processes under irrigation. Soil Science, No.10, 1987, article 65.

72.Soils of the Republic of Karakalpakstan (Shumanai district), book 2, IPA AS RUz, 1995. 113 c.

73. Prikhodko V.E. Degradation of steppe soils under irrigation// Theses of reports of the 111th Congress of Dokuchaevsky society of soil scientists (July 11-15, 2000, Suzdal) Book 2. Москва,2000,стр.242-243.

74. Prokopov D.N. Proper crop rotations and rational use of lands. Minsk. Nauka i tekhnika. 1976. 188c.

75.Rajabov A. I., Gafurova N. Influence of organic matter on cation exchange capacity and accumulation of heavy metals in soil and tobacco plants // Esh kimyogar (ilmiy makrlalar tuplami).- Samarkand, 2001.- No. 3. 108-109 c.

76.Razakov R.M. Problems of water resources and environmental risk affecting the health of the population of Priaralie. Materials of the international seminar "Ecological factors and maternal/child health in the region of the Aral Sea crisis" Izd. R Uz. 2001, C. 20-27.

77.Riskieva H.T., Toshkenboev O. Change of humus state of meadow soils of desert zone of Zarafshan valley under the influence of irrigation // Bulletin of agrarian science of Uzbekistan, № 1(7),2002, p.69-72.

78.Ryzhov S.N. To the question about preparation of carbonate soils for mechanical analysis. Bulletin of SoyuzNIIKhI, No. 7, 1935.

79.Ryzhov S.N., Ziyamukhamedov I.A. Changes in the content and composition of humus of gray soils and takyr soils under irrigation - "Soil Science", 1971, № 7.

80.Ryzhov S.N., Tashkuziev M.M. Change of chemical composition of sulfur soils and takyr soils under the influence of irrigation and high culture of farming. In the collection Irrigated soils and methods of their study. Tashkent, 1976.

81.Salokhitdinov Z., Usmonov R., Dustmukhamedov Turok unumdorligi - mul khosil mezoni // Ozbekiston kishlok khuzhaligi.2003, № 5, 28 b.

82. Sattarov J.S., Toshmatov N.I. Technology of compost preparation from different wastes and their influence on soil and cotton yield. 1993. No.2, pp.24-27.
83.Sattarov J.S., Kazak O.R. Microelements in soils of the belt of typical sierozem depending on their agrotechnical condition and humus content // Bulletin of Agrarian Science of Uzbekistan, № 2,2000, p.52-53.
84.Sattorov D., Kuziev R., Karimberdiyeva A. Changes in soil fertility under the influence of irrigation //Agriculture of Uzbekistan, No. 6 2001, 22-23 pp.
85.Tashkuziev M.M. Humus state of the main soils of the lower part of the Amudarya delta subjected to desertification processes. TuproTsdan o Pilona foidalanishning ekologik zhihatlari. Ilmiy-amaliy conf. mat. Toshkent, 1977, p. 279-281.
86.Tashkuziev M.M., Ziyamukhamedov I.A. Moddalarning biologic ailanishini tuproklar kimiyoviy tarkibini shaklanishdadagim ahamiyati. Er resurslariridan okilona foidalanish va tuproklarni mukhofazalash. Ilimiy Tuplam. Toshkent.2001.
87.Toshkuziev M.M. Uzbekistan tuproklari humusi va unin tuprok unimdorligida ahamiyati. Tuprokshunoslar va agrokiyogarlar 111-chi kuroltoyida maruzalari va tezislari. Toshkent.2000.
88.Toshmatov N.I. Influence of composts made from household and industrial wastes on agrochemical properties of soil and cotton yield. Author's abstract of dissertation for the degree of candidate of agricultural sciences.Tashkent, 1996, 21 p.
89.Trushin V.F. Rye crops in the Sverdlovsk region // Proceedings of the Sverdlovsk Agricultural Institute. Vyp. XУ. 1969. C.43-47.
90.Turapov I. Thermal regime of soils of vertical zonality of Western Tien-Shan and ways of its regulation. Tashkent, 1994, p. 47.
91. Tyulin T.V. Dynamics of water resistance of soil structure on virgin lands of Kazakhstan in connection with their cultivation // V sb. of works on agronomic physics. Vyp.8. 1960. C.170.
92. Fursov V.N. Agrochemical and biological properties of soils of different agrofonts and their change under the influence of crop rotations and long-term application of fertilizers - In the collection: Research works on cotton growing. Ashgabat, "Turkmenistan", 1966.
93.Khalikulov Sh.T. Optimization of properties and regimes of irrigated soils of gray soil belt by mulching and compost application during cotton cultivation. Tashkent, 1996, p. 48.
94.Shishov L.L., Durmanov D.N. Modern Concepts of Soil Fertility Management // In Sb. Soil Fertility, Problems, Research, Model. Dokuchaev Soil Institute. Dokuchaev Soil Institute. M.,1985. C.3-18.
95.Shokirov A. Eggplant usimligining khasildorligiga ekish schemesining tasiri. \\ Bulletin of Agrarian Science of Uzbekistan, № 3(5), 2001, p. 114-115.
96. Shchapova L. N. Microbiological aspects of transformation of organic matter with different carbon content to nitrogen // Theses of reports of the 111th Congress of the Dokuchaevsky Society of Soil Scientists (July 11-15, 2000, Suzdal) Book 2. Москва,2000,стр.68-69.
97.Yakubov M., Kudratov U. Ameliorative-ecological processes on irrigated massifs on the

right bank of the Amudarya River // Agriculture of Uzbekistan.No.4, 2001, pp.35-37.
98.Buol S.W., Hole E.D., McCracken R.J.. *Soil Genesis and classification* // Second Edition. The Iowa State University Press, Ames, 1980. 406p.
99.Carol Cox, and staff: Booklet 26: Learning to Grow All Your Own Food. Willits. CA: Ecology Action : 1991.
100. Henry. Dr. Foth. D. "A Study Soil Science" Copyright @ April, 1970. LaMotte Company PO Box 329, Chestertown, Maryland 21620 U.S.A.
101. Jenny, Hans. The Soil Resource: Origin and Behavior. New York: SpringerVerlag, 1980.377 pp.
102. John Jeavons John Jeavons. Booklet 14: The Complete 21-Bed Mini-Farm. Willits CA: Ecology Action 1991.
103. John Jeavons J. Mogador Griffin, and Robin Letter. The Backyard Homestead Willits CA: Ecology Action, 1983.
104. John Jeavons How to Grow More Vegetables.1995.208 pp.
105. John JeavonsDavid. Duhon, and Gebhard Cindy. One Circle. Ecology Action, 1984.
106. Kusselevski Z. Influence of organic and mineral fertilizers on the humus content in the soil // Zemdeloje, 1973, No. 5(2), P.322.
107. McLeod, Edwin. Feed the Soil. Organic Agriculture Research Institute (P.O. Box 475, Grator CA 95444),1982. 209 pp.
108. Richardson G., Mueller-Beilschmidt P. *Winning with water* // Soil- moisture monitoring for efficient irrigation. INFORM, Inc. 1988, 173pp.
109. Shexson. T.F. Land Husbandry: A Framework for Soil and Water Conservation. Soil and Water Conservation Society (7515 Northeast Ankeny Road, Ankeny, IA 50021-9764), 1989. 64 pp
110. Zhollybekov B. *Alteration of Soils in the irrigated Zone of AmuDarya river delta //* Eurasian Soil Science, 28(10), 1996, 265-274p.p.
111. Jollibekov B. Irrigation Practices in Uzbekstan // 3rd International Conference on Efficient Use and Management of Water. Efficient 2005 - Santiago-Chile. 1- 4p.p.

appendix

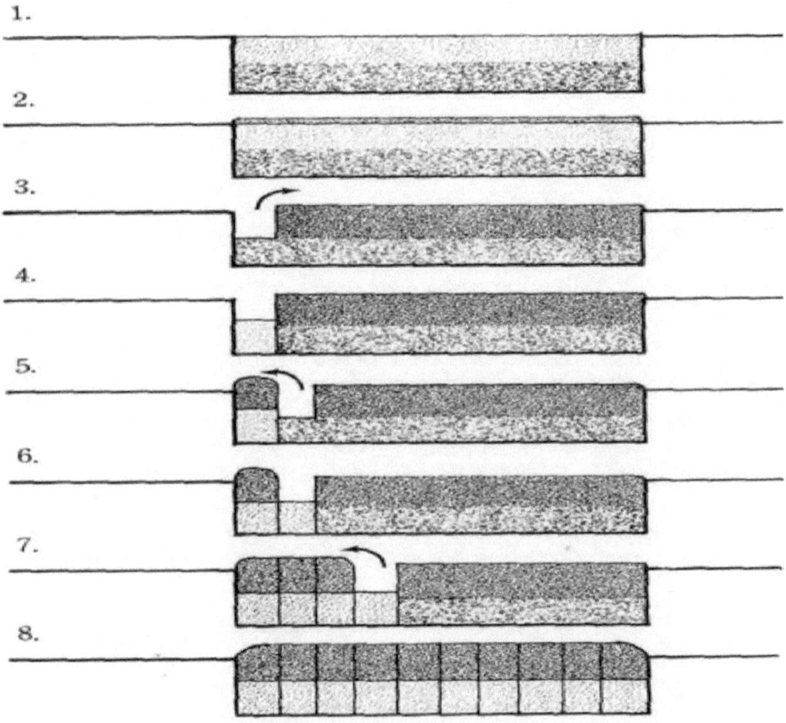

Fig 1 . Method of preparation of "Double digging".

A

Б

Fig 2. Compost heap according to D. Jevons (A) and B. Jolybekov (B).

Figure 3. Compost ready for use.

A

Б

Fig 4. Cultivation of sidedress (A) and its fallowing in spring with a mini-tractor (B)

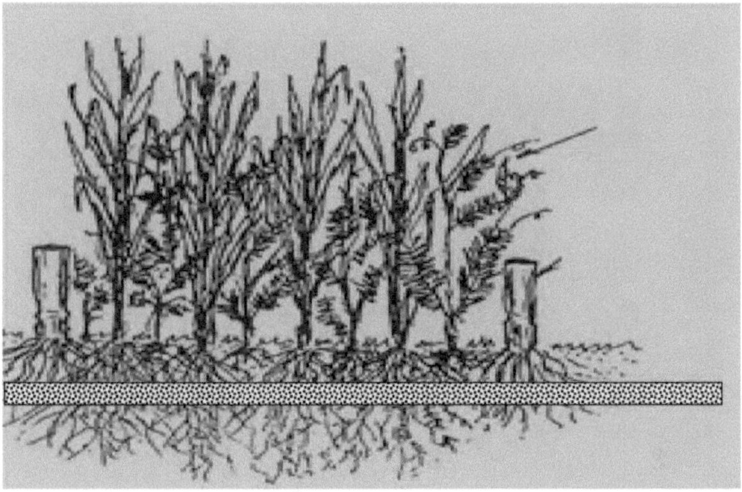

A

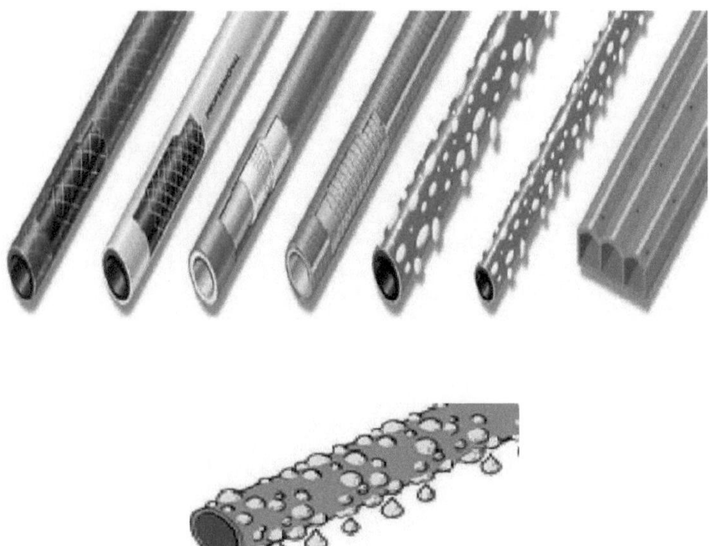

Б

Figure 5. Laying of impregnating hose (A) and its molds (B).

А

Б

Fig 6. Soaking hose, planting of eggplant (A) and tomato (B) seedlings.

Fig 7 . Installation of plastic bottles (A) and planting of tomato seedlings (B).

Б

Fig 8. Phenological observations of growth and development of eggplant (A) and tomato (B).

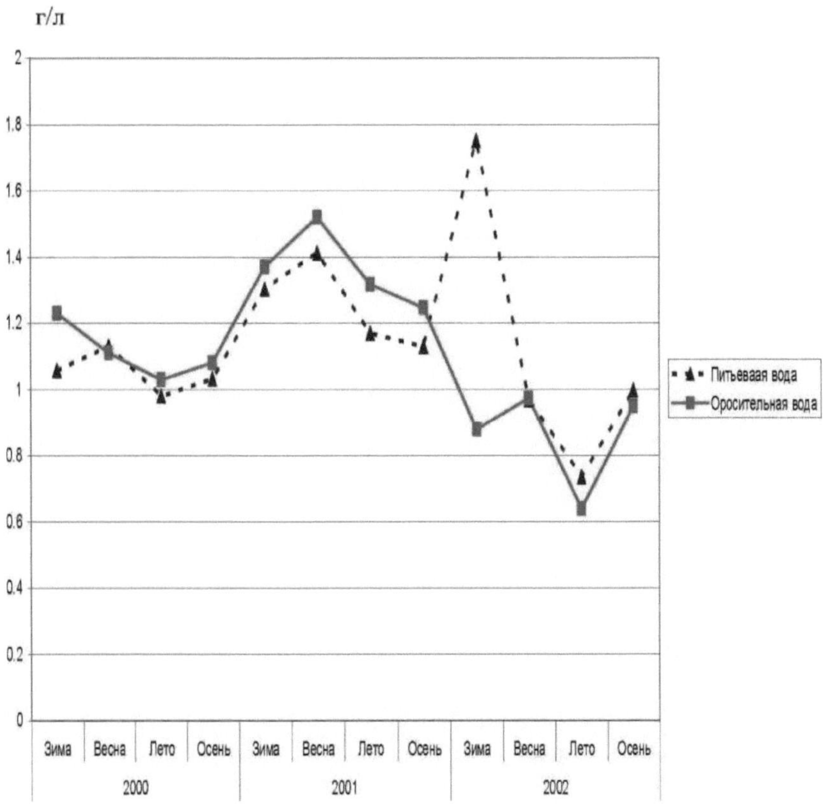

Figure 9. Annual and seasonal variation of water-soluble salts content (g/l).

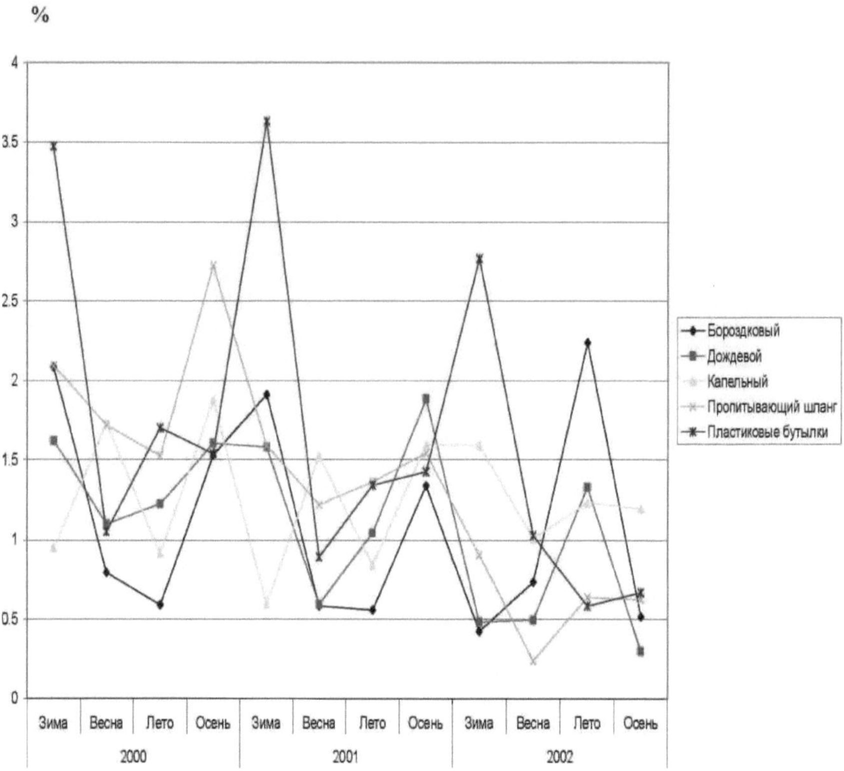

Fig10. Seasonal and annual changes in water-soluble salt content (%) in 0-30 cm layer under different irrigation methods.

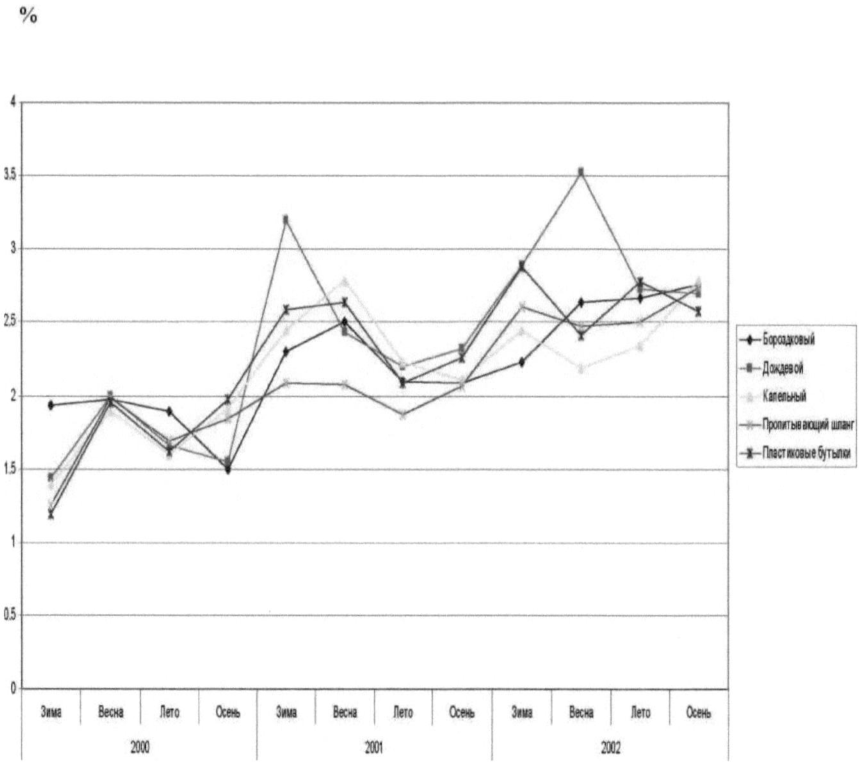

Fig. 11: Seasonal and annual changes in humus content in 0-30 cm layer under different irrigation methods (%).

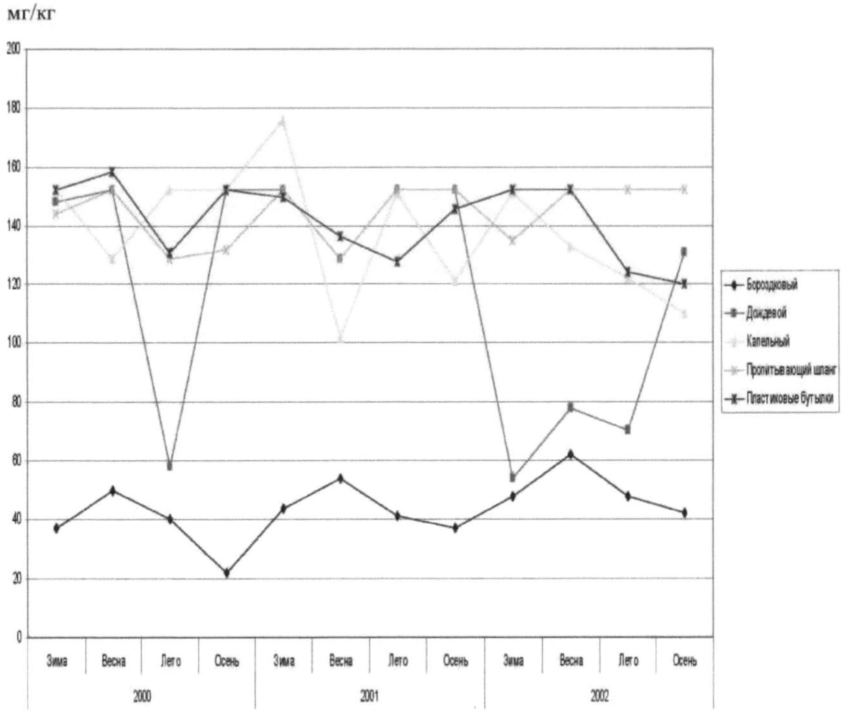

Fig. 12: Seasonal and annual changes in nitrate nitrogen content, in 0-30 cm layer under different irrigation methods (mg/kg).

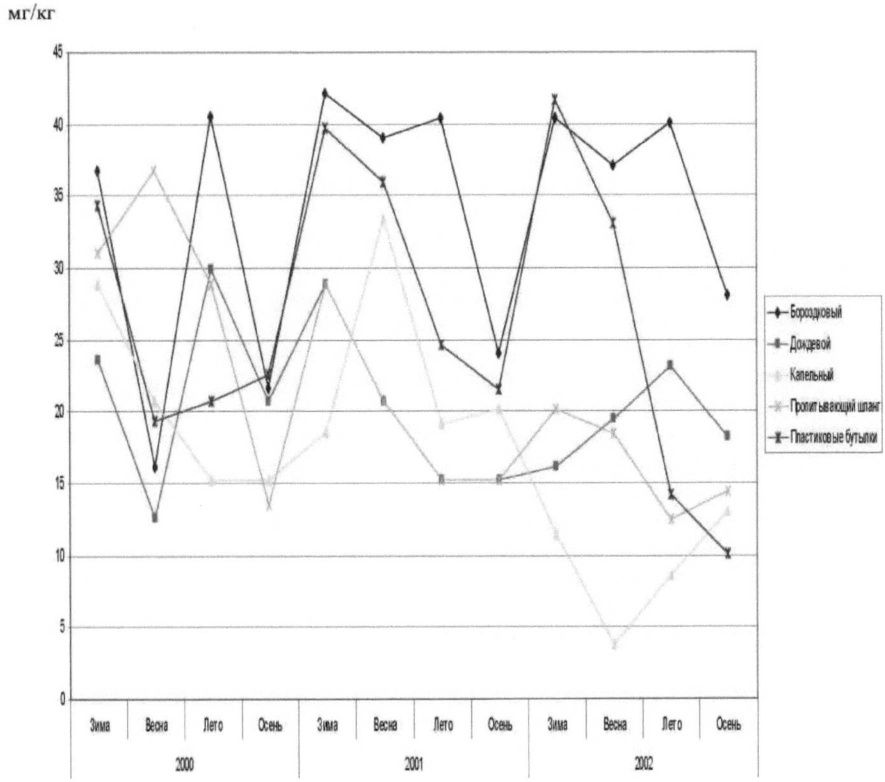

Fig. 13: Seasonal and annual changes in mobile phosphorus content, in 0-30 cm layer under different irrigation methods (mg/kg).

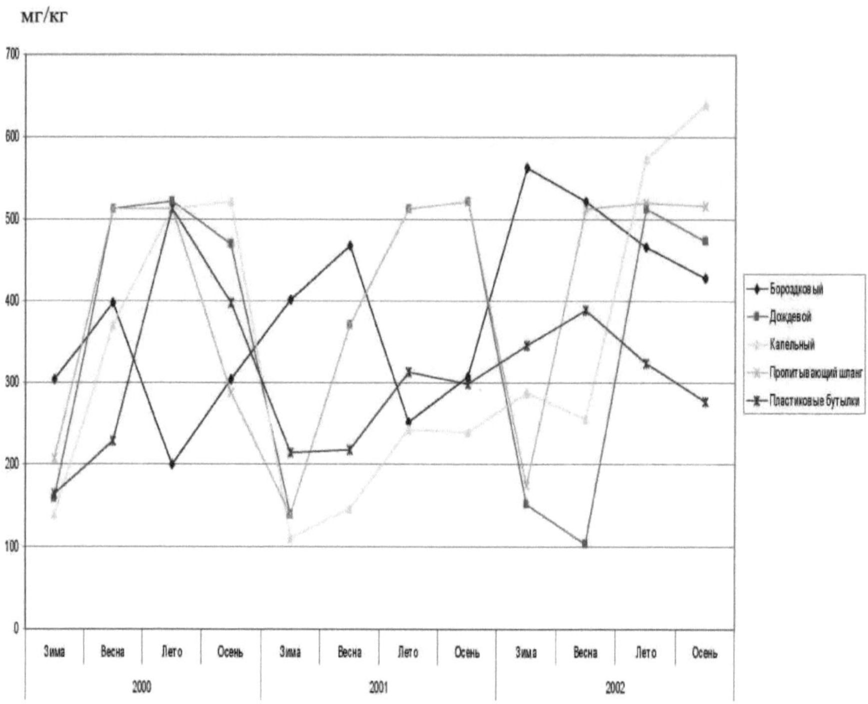

Fig. 14: Seasonal and annual changes in the content of mobile potassium, in the 0-30 cm layer under different irrigation methods (mg/kg).

MIX
Papier aus verantwortungsvollen Quellen
Paper from responsible sources
FSC® C105338

Printed by Books on Demand GmbH, Norderstedt / Germany